advancing learning, changing lives

Students' Book
Extension Units

For GCSE Biology, GCSE Chemistry and GCSE Physics

Edexcel 360Science

Edexcel's own course for the new specification

David Applin
Gerry Blake
Iain Brand
Michael Brookman
Steve Gray

A PEARSON COMPANY

How to use this book

The book is divided into 6 topics. Each topic has a one-page introduction and is then divided into sixteen double page sections. At the end of each topic there is a set of questions that will help you practise for your exams and a glossary of key words.

As well as the paper version of the book there is a CD-ROM called an ActiveBook. For more information on the ActiveBook please see the next two pages.

What to look for on the pages of this book:

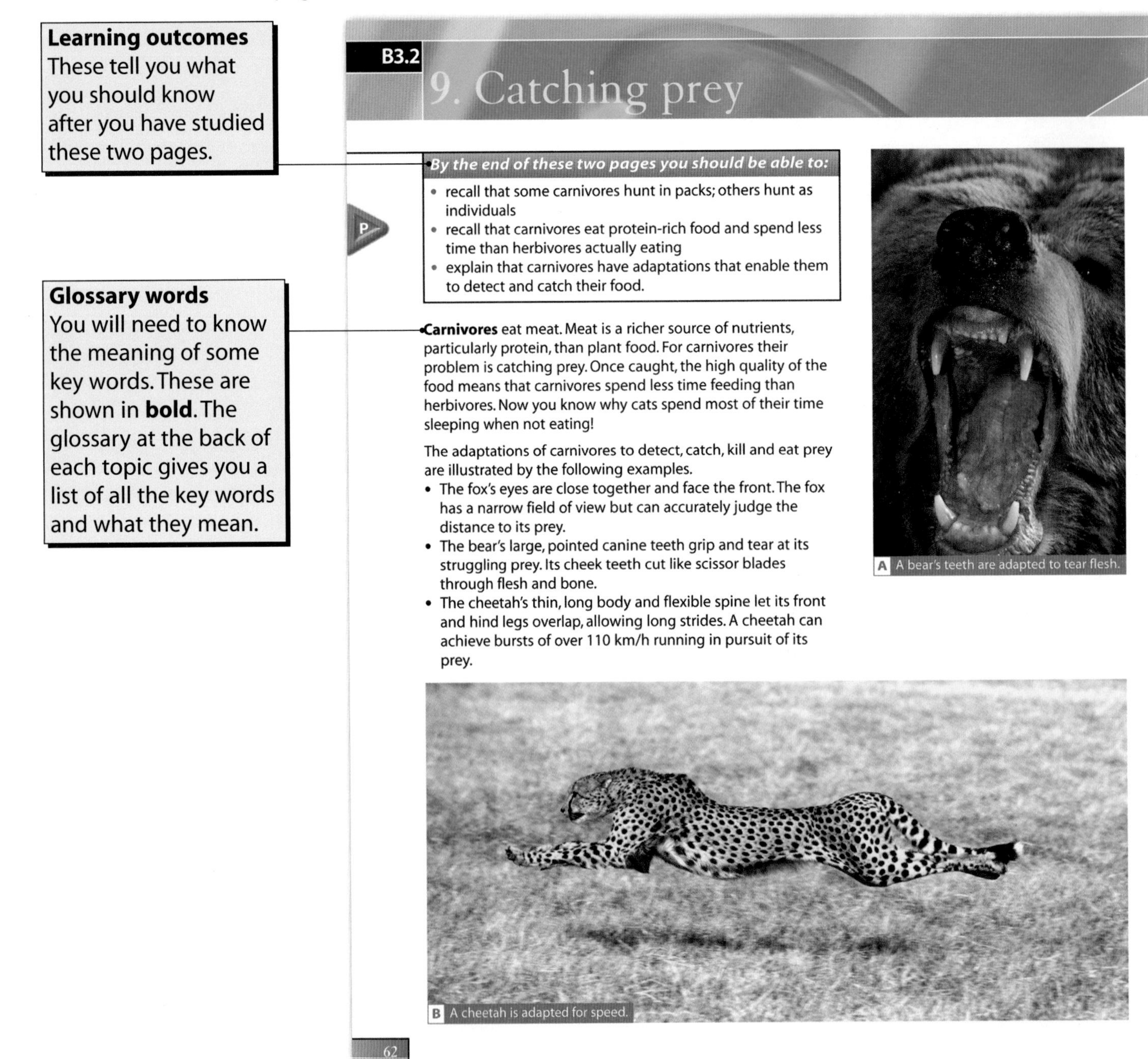

Learning outcomes
These tell you what you should know after you have studied these two pages.

Glossary words
You will need to know the meaning of some key words. These are shown in **bold**. The glossary at the back of each topic gives you a list of all the key words and what they mean.

B3.2

9. Catching prey

By the end of these two pages you should be able to:

- recall that some carnivores hunt in packs; others hunt as individuals
- recall that carnivores eat protein-rich food and spend less time than herbivores actually eating
- explain that carnivores have adaptations that enable them to detect and catch their food.

Carnivores eat meat. Meat is a richer source of nutrients, particularly protein, than plant food. For carnivores their problem is catching prey. Once caught, the high quality of the food means that carnivores spend less time feeding than herbivores. Now you know why cats spend most of their time sleeping when not eating!

The adaptations of carnivores to detect, catch, kill and eat prey are illustrated by the following examples.

- The fox's eyes are close together and face the front. The fox has a narrow field of view but can accurately judge the distance to its prey.
- The bear's large, pointed canine teeth grip and tear at its struggling prey. Its cheek teeth cut like scissor blades through flesh and bone.
- The cheetah's thin, long body and flexible spine let its front and hind legs overlap, allowing long strides. A cheetah can achieve bursts of over 110 km/h running in pursuit of its prey.

A A bear's teeth are adapted to tear flesh.

B A cheetah is adapted for speed.

62

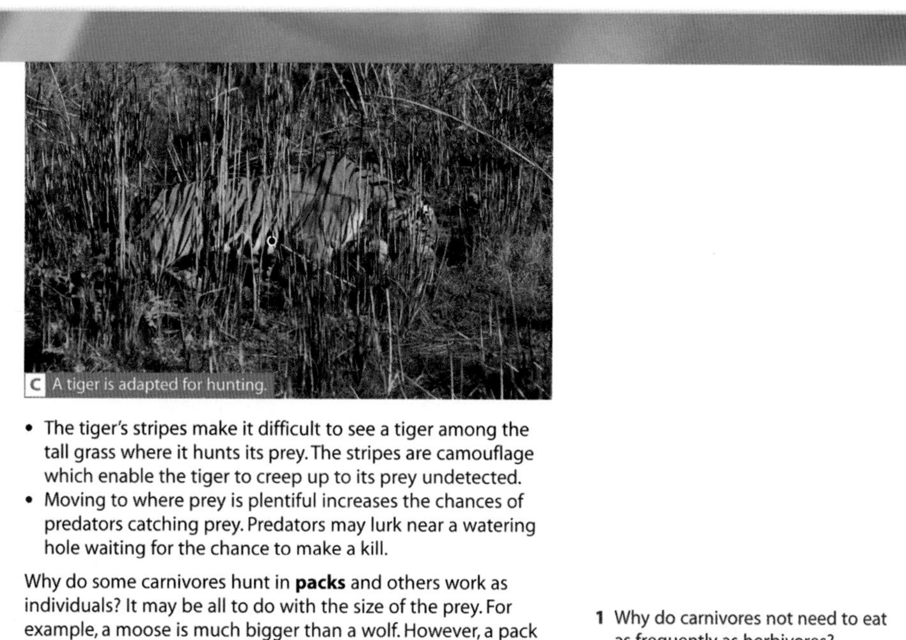
C A tiger is adapted for hunting.

- The tiger's stripes make it difficult to see a tiger among the tall grass where it hunts its prey. The stripes are camouflage which enable the tiger to creep up to its prey undetected.
- Moving to where prey is plentiful increases the chances of predators catching prey. Predators may lurk near a watering hole waiting for the chance to make a kill.

Why do some carnivores hunt in **packs** and others work as individuals? It may be all to do with the size of the prey. For example, a moose is much bigger than a wolf. However, a pack of wolves working together can bring down a moose where a wolf on its own would probably fail.

Feeding in a pack means that there are fewer leftovers from the kill. Scavengers are animals that feed on the remains of prey killed by predators. Packs of carnivores are able to chase off scavengers. This means that each member of the pack has more meat for it to feed on than an animal hunting alone.

Have you ever wondered?

Why do dogs greet each other by sniffing?

Most wild cats hunt alone for small prey, leaving few leftovers for scavengers. The hunter eats most of the prey it kills and so pack life would have few advantages.

3 What is the advantage of hunting in packs?

4 What are the advantages of scavenging for food?

5 Use examples to show how hunting in packs and hunting alone are equally successful strategies in different species of carnivores.

Have you ever wondered?

Why are dogs so different from cats?

1 Why do carnivores not need to eat as frequently as herbivores?

2 Describe how the following characteristics help predators to catch, kill and eat prey:
a eyes
b legs
c teeth
d colouring
e hunting near watering holes.

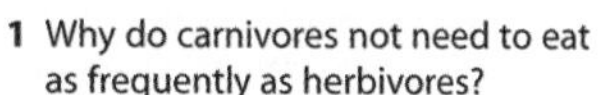

Summary Exercise | Higher Questions | Extension Questions

A P ? P ?

'Have You Ever Wondered?'
These questions are there to help you think about the way science works in your life. Your teacher might ask you what you think.

Questions
There are lots of questions on the page to help you think about the main points in each double-page section.

How to use your ActiveBook

The ActiveBook is an electronic copy of the book, which you can use on a compatible computer. The CD-ROM will only play while the disc is in the computer. The ActiveBook has these features:

Glossary
Click this tab to see all of the key words and what they mean. Click 'play' to listen to someone read them out to help you pronounce them.

DigiList
Click on this tab to see menus which list all the electronic files on the ActiveBook.

ActiveBook tab
Click this tab at the top of the screen to access the electronic version of the book.

Key words
Click on any of the words in **bold** to see a box with the word and what it means. Click 'play' to listen to someone read it out for you to help you pronounce it.

Interactive view
Click this button to see all the bits on the page that link to electronic files, like documents and spreadsheets. You have access to all of the features that are useful for you to use at home on your own. If you don't want to see these links you can return to **Book view**.

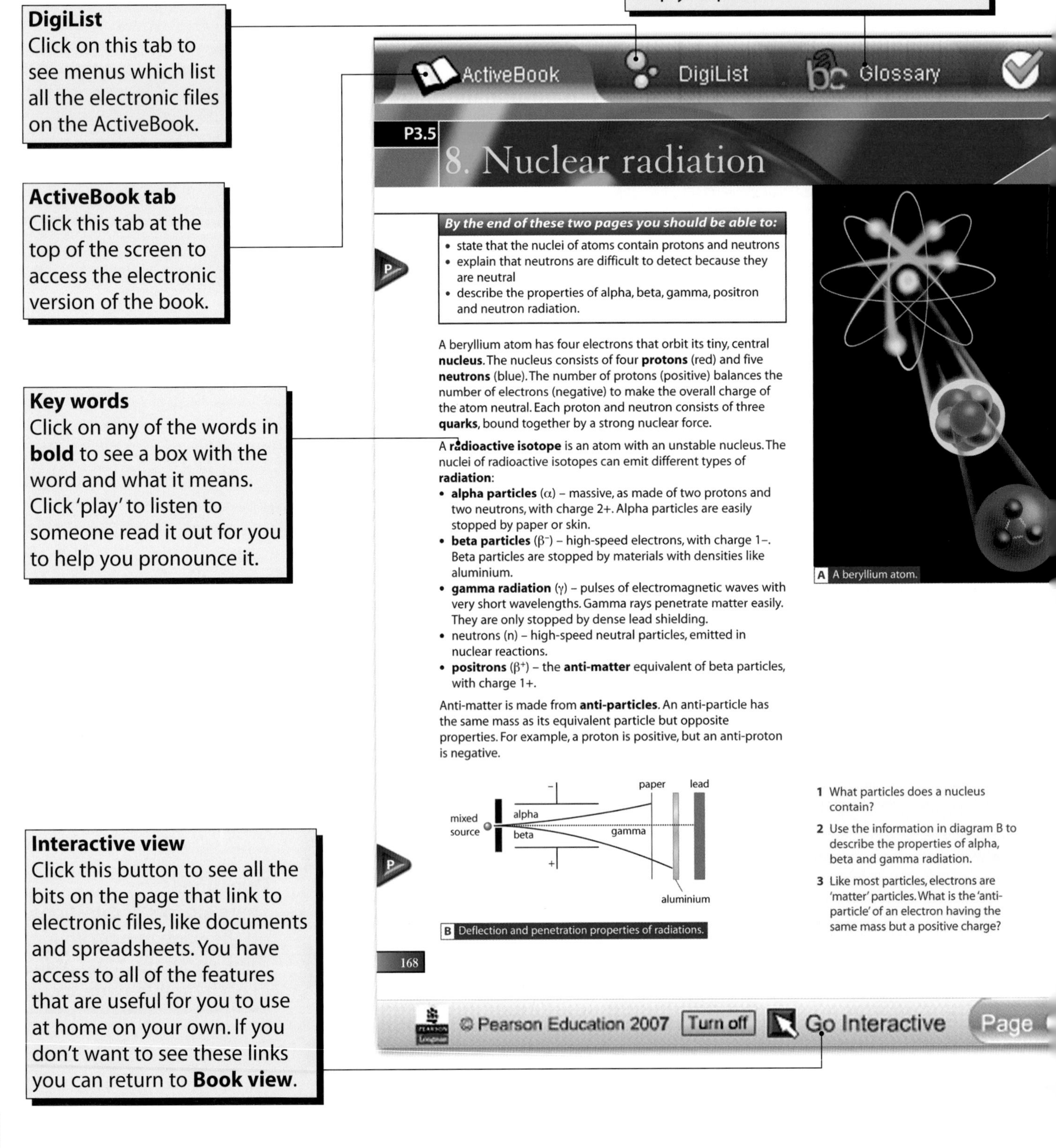

Target sheets
Click on this tab at the top of the screen to see a target sheet for each topic. Save the target sheet on your computer and you can fill it in on screen. At the end of the topic you can update the sheet to see how much you have learned.

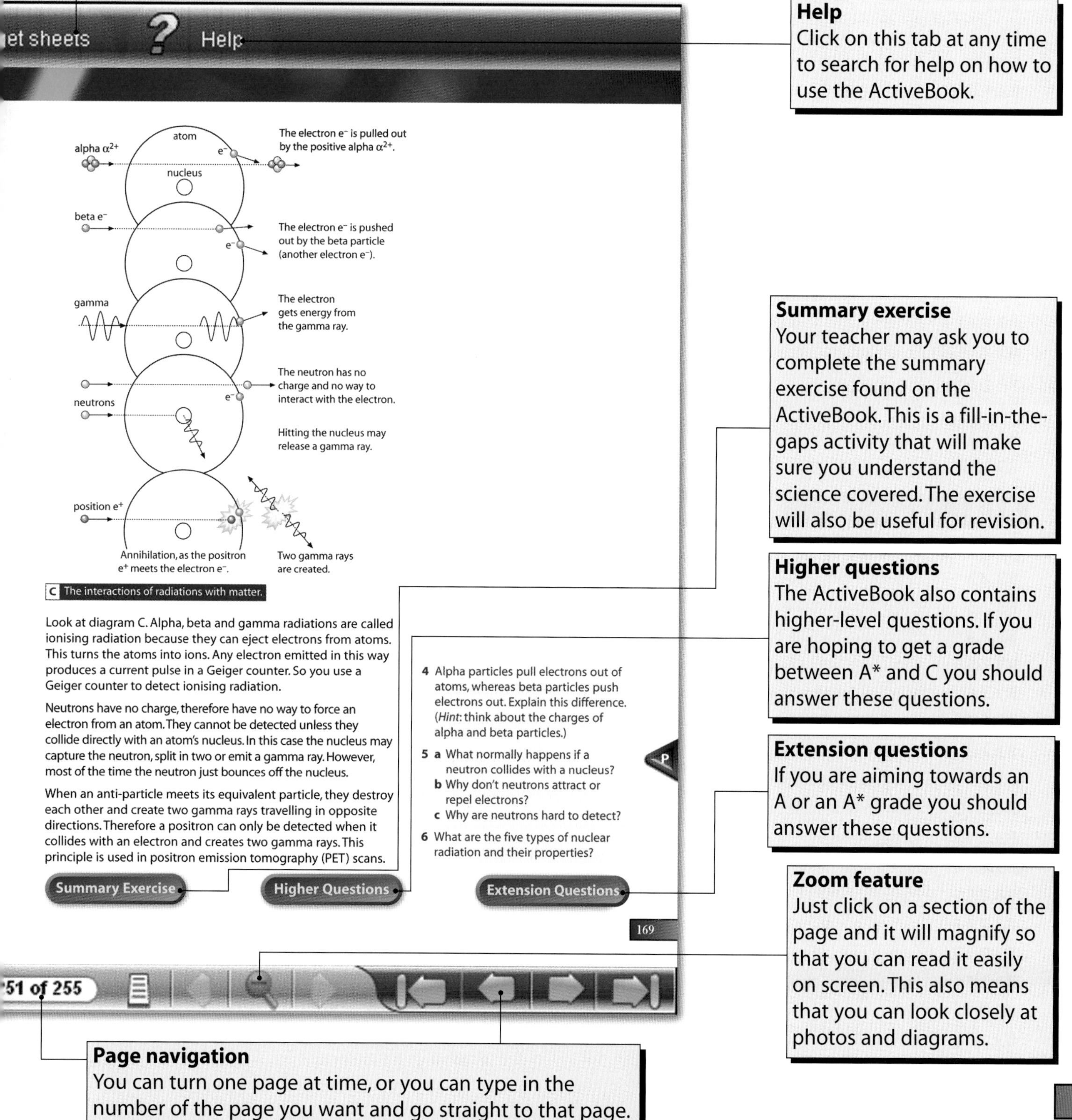

Help
Click on this tab at any time to search for help on how to use the ActiveBook.

Summary exercise
Your teacher may ask you to complete the summary exercise found on the ActiveBook. This is a fill-in-the-gaps activity that will make sure you understand the science covered. The exercise will also be useful for revision.

Higher questions
The ActiveBook also contains higher-level questions. If you are hoping to get a grade between A* and C you should answer these questions.

Extension questions
If you are aiming towards an A or an A* grade you should answer these questions.

Zoom feature
Just click on a section of the page and it will magnify so that you can read it easily on screen. This also means that you can look closely at photos and diagrams.

Page navigation
You can turn one page at time, or you can type in the number of the page you want and go straight to that page.

Contents

B3.1

Biotechnology

A Using biotechnology to make yoghurt.

Biotechnology is the scientific word for the way we use plant cells, animal cells and **microorganisms** to produce substances useful to us or to carry out useful processes. We can use biotechnology to produce food and drink. Yeast allows us to make wine, beer and bread, and lactic acid bacteria turn milk into yoghurt and cheese.

Many of the drugs used today are sourced from plants. Biotechnology allows us to develop new plant-based drugs. Organisms used in biotechnology can also be genetically modified to produce new medicines and other useful substances. However, new technology can raise ethical questions.

In this topic you will learn that:

- the food industry has traditionally made much use of biotechnology in the production of many food items, for example cheese, yoghurt, alcohol, chocolate, soy sauce and, more recently, mycoproteins and prebiotics
- plants can be modified to be resistant to herbicides and/or pests and this has environmental implications
- the pharmaceutical industry generates a lot of money annually and consideration of the contributors to this profit and its distribution is needed
- stem cell research must consider many ethical questions, including the definition of 'life'
- organisms can be genetically modified to produce substances, including medicines, that are of direct use to human health.

Look at these statements and sort them into the following categories:

I agree, I disagree, I want to find out more

- Eating GM food is not natural.
- Knowing about the human genome might affect human rights.
- Enzymes are protein catalysts which speed up chemical reactions taking place in our cells.
- Stem cell research may lead to the development of treatments for different diseases.
- Developing biotechnology is the only way we can solve the worldwide food shortage.
- If I eat what I like my diet will be well-balanced.

V

D

1. Functional foods

By the end of these two pages you should be able to:

- recall that some functional foods are produced by fermentation; others are not
- recall that prebiotics are functional foods which are marketed as providing health benefits
- recall that prebiotics contain oligosaccharides
- recall that plant stanol esters in 'spreads' lower cholesterol levels in the blood.

Functional foods contain an added ingredient not normally found in that food. This ingredient is supposed to promote good health. In other words functional foods contribute to our wellbeing as well as keeping us alive.

Each of us carries about 1 kg of different types of **bacteria** in our gut. Bacteria are microorganisms. For most of the time the numbers of bacteria in our gut are in balance. Poor diet, stress, food poisoning and the use of antibiotics, for example, can disturb the balance. The numbers of some types of bacteria can grow at the expense of the others and diseases of the gut can develop.

We sometimes call troublesome bacteria 'bad' and the others 'good'. 'Good' bacteria suppress the activities of the 'bad'. Our better understanding of the way the different types of bacteria affect each other in the gut has led to the development of functional foods called **prebiotics** and probiotics. These foods boost the numbers of 'good' bacteria over the 'bad'.

1 What is a functional food?

2 What are the differences between prebiotic food and probiotic food?

Prebiotics contain added sugars called **oligosaccharides**. When we eat these foods the sugars are not digested by the body but are food for the 'good' bacteria in our gut. The bacteria increase in numbers. Probiotics contain the 'good' bacteria themselves. When we eat these foods the bacteria lost because of poor diet, stress and so on, are replaced. Encouraging growth of the 'good' bacteria in our gut and replacing those lost may help us to keep well. Probiotics are produced by **fermentation**, when bacteria ferment milk and other foods. Prebiotics are not produced by fermentation.

A Probiotic foods.

The spreading margarine Benecol® is a functional food. It contains stanol **ester**. This ingredient is added to the spread. It comes from plant sterols which are fatty substances found naturally in foods like wheat and maize.

Studies have shown that people who include plant stanols in their diet for a year might expect the **cholesterol** levels in their blood to fall by up to 10%. Raised levels of cholesterol in the blood increase a person's risk of developing heart disease. Spreading Benecol® therefore reduces the risk of developing heart disease – so its makers claim.

Do functional foods improve our health? Manufacturers say that the foods prevent, treat or cure disease. However, the difference between our hopes and manufacturers' claims is the reason why there are doubts about the value of eating functional foods as a way of staying healthy. Because of the uncertainty, many scientists think that the foods should be tested in the same way that new drugs are tested before the drugs can be used to treat diseases.

At the moment it seems best to be cautious. A balanced diet provides many of the ingredients added to functional foods. So, it is important to think carefully about what you eat before rushing out and spending money unnecessarily.

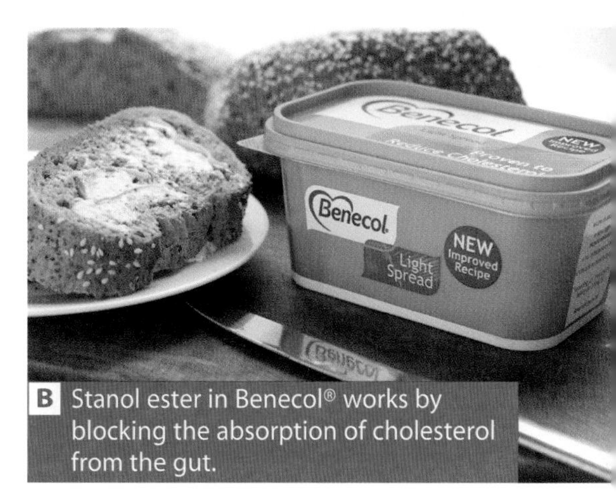

B Stanol ester in Benecol® works by blocking the absorption of cholesterol from the gut.

3 Explain the role of oligosaccharides in prebiotic food.

4 How do foods which contain plant stanols reduce the risk of a person developing heart disease?

5 Give your opinion on the role of functional foods in a balanced diet.

Summary Exercise

Higher Questions

Extension Questions

2. Making yoghurt and soy sauce

By the end of these two pages you should be able to:

- describe the production of yoghurt from milk as the conversion of lactose to lactic acid using bacteria
- explain that the commercial production of soy sauce includes fermentation of a mixture of cooked soya beans and roasted wheat using *Aspergillus*, further fermentation using yeasts and then *Lactobacillus*, filtration, pasteurisation and sterile bottling.

Milk is a nutritious food but soon turns sour unless kept in a refrigerator. Even then refrigerated milk will only keep for a few days. Long before refrigerators we knew that lactic acid bacteria could be used to preserve milk. The bacteria turned milk into cheese or yoghurt which keep for much longer than milk itself.

A A starter culture of bacteria is added to the milk.

To make yoghurt, milk is heated to sterilise it, killing any bacteria already present. Then the milk is stirred thoroughly to mix up its ingredients and milk protein is added. A starter culture of *Lactobacillus* bacteria is added to the milk while it is still warm. The *Lactobacillus* bacteria ferment the sugar **lactose** in the milk to **lactic acid**

$$\text{lactose} \xrightarrow{\textit{Lactobacillus}} \text{lactic acid}$$

The lactic acid lowers the pH of the milk (the milk becomes acid) to the point where the milk proteins coagulate (solidify). The semi-solid milk is raw yoghurt.

The raw yoghurt is cooled quickly. Different flavourings of fruit may be added to the yoghurt or it may be put into pots as natural yoghurt. The pots are then sealed and ready for sale.

1 What is fermentation?

2 Name the type of bacterium used to make yoghurt.

3 Why is the milk used to make yoghurt:
a sterilised
b stirred thoroughly
c kept warm when the bacteria are added?

4 Why is yoghurt kept cool once it's made?

Soy sauce is a fermented sauce made from soya beans. It originally came from China and is an important flavouring used in cooking. Producing soy sauce depends on the activities of *Aspergillus* mould, *Lactobacillus* bacteria and yeast.

B Soy sauce.

Soya beans are cooked, which kills all of the bacteria on their surface, and mixed with ground roasted wheat. *Aspergillus* mould is added to the mixture. The mixture is spread out on warm shallow trays and supplied with air. **Enzymes** produced by the *Aspergillus* mould catalyse the breakdown of the proteins and carbohydrates in the mixture. The **amino acid** and sugar content of the mixture increases.

Brine (sodium chloride solution) is added to the mixture. It gives soy sauce its salty taste and helps to preserve the final product. Yeasts and *Lactobacillus* bacteria, which are able to tolerate high levels of sodium chloride and low levels of oxygen, are added to the brine mixture. The conditions stop the activities of the *Aspergillus* mould. The sugars in the mixture are fermented by the yeasts and *Lactobacillus* bacteria.

Raw soy sauce is drained from the mixture. The liquid is filtered and cleared of any sediment (**filtration**). The filtered and cleared liquid is heated to 72 °C (**pasteurisation**) and stored to allow its flavours to develop. It is then put into sterilised bottles ready for sale.

5 Name three types of microorganism used to produce soy sauce.

6 Why is soy sauce:
 a filtered
 b pasteurised
 c stored before it is sold?

7 Draw flow diagrams to show the process of:
 a yoghurt production
 b soy sauce production.

Summary Exercise

Higher Questions

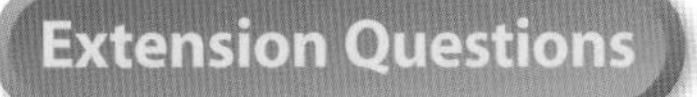

3. More on microorganisms and food

By the end of these two pages you should be able to:

- describe the production of the enzyme chymosin, produced by genetically altered microorganisms
- recall that chymosin is used in the manufacture of vegetarian cheese
- describe microbial products used in food, including vitamin C, carrageen extract, enzymes such as invertase, citric acid, amino acids such as glutamic acid, monosodium glutamate.

Rennet is vital to the process of cheese making. It is a mixture of two protein-digesting enzymes. These enzymes cause the milk to form solid curds which separate from the liquid, which is called whey. The curds are raw cheese. Traditionally rennet comes from the stomach of slaughtered calves.

Rennet consists mostly of the enzyme **chymosin**. We can now use genetically modified (GM) bacteria to produce chymosin. The enzyme works in exactly the same way as calf rennet but its activity is more predictable, it contains fewer impurities and is acceptable to vegetarians.

The cheese made using GM chymosin is itself rated GM-free because it is not made using GM organisms, even though the enzyme used in its production is made by GM organisms.

Gelling agents are substances added to food to thicken it and make it less runny. Ice creams, jellies, milk shakes and frozen desserts are just some of the foods which contain gelling agents. Gelatine made from animal bones is an everyday example of a gelling agent.

Foods containing gelatine are unacceptable to vegetarians because the gelatine comes from animals. However, various alternatives which do not come from animals are available. Carrageen extract, which comes from the reddish-purple seaweed called carrageen, is an example.

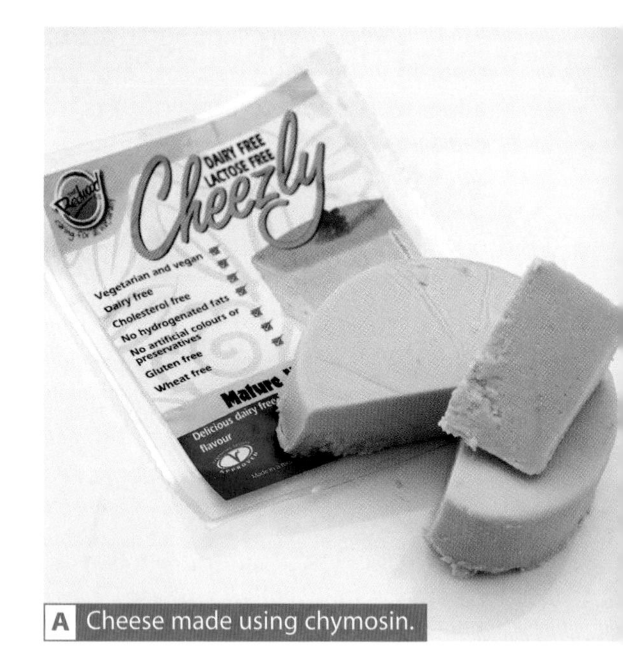

A Cheese made using chymosin.

1. Name the enzyme traditionally used in cheese making. How is the enzyme obtained?
2. What are GM organisms?
3. Why is cheese made using the enzyme chymosin produced by genetically modified (GM) microorganisms rated as GM-free?
4. Why is gelatine not acceptable to vegetarians, but carrageen extract is?

The mould fungus *Aspergillus niger* is used to produce **citric acid**. When citric acid is added to fizzy drinks, sweets, jams and jellies, it enhances (brings out) the flavour of the foods. It also prevents the breakdown of vitamin C in food.

B These foods are made using microorganisms.

The bacterium *Corynebacterium glutamicum* is used to produce the amino acid **glutamic acid**. Neutralisation of the acid makes a sodium salt called monosodium glutamate (MSG). When added to food, MSG enhances the food's flavour. It is particularly popular in Chinese food as a flavour enhancer.

Different species of the bacterium *Acetobacter* are used to produce vitamin C. The process is a fermentation in two stages starting with glucose.

The **yeast** fungus *Saccharomyces cerevisiae* is used to produce **invertase** (also called sucrase). The enzyme promotes the breakdown of the sugar sucrose into glucose and fructose.

$$\text{sucrose} \xrightarrow{\textit{invertase}} \text{glucose} + \text{fructose}$$

Sweet-makers use invertase to convert the solid insides of chocolate into a liquid centre. Glucose isomerase is another enzyme used in the sweet-making industry. This enzyme converts glucose into fructose, which is sweeter. Fructose syrups which are used to make sweets, are very sweet indeed!

5 What is MSG and what is it used for?

6 What type of bacterium is used to make:
a MSG
b vitamin C?

7 Name the microorganism used to make citric acid.

8 Why is citric acid added to food and drink?

9 Why is fructose often used by the sweet-making industry, rather than glucose?

10 Describe the importance of microorganisms in food production.

Summary Exercise | Higher Questions | Extension Questions

4. Eating well

By the end of these two pages you should be able to:

- describe the importance of having a well-balanced diet
- explain that obesity may lead to a number of health problems.

Our diet is the food we eat and the liquids we drink. Food and drink contain nutrients: carbohydrates, fats, oils, proteins, vitamins and minerals. Food also contains fibre, water and other substances which make it tasty and colourful. The nutrients, fibre and water in our food and drink should be in the correct amounts and proportions for good health. If they are, then we have a balanced diet.

If a diet consists of a single food then that food will be unhealthy, whatever it is. No single food contains all the nutrients in the proportions we need for long-term good health. For example, beef and wholemeal bread lack vitamins A, C and D and are low in calcium. Beef also lacks fibre, which wholemeal bread provides. Together beef and wholemeal bread provide more nutrients than either on its own. Between them, however, vitamins A, C and D are still missing. Add salad, fruit and vegetables, and vitamins A and C are brought into the diet. Milk and cheese provide more calcium and add vitamin D. The mixture of foods provides enough nutrients, fibre and water for a healthy, balanced diet. On its own, each food lacks something which all the foods together make up between them. Each of these foods – beef, wholemeal bread, fruit and vegetables, milk and cheese – belongs to a different food group.

1 What is meant when a diet is described as being balanced?

2 What are the basic four food groups?

A The four food groups. You should eat a variety of foods from each group every day.

The idea of food groups has been developed to help us choose a balanced diet. However, we should also aim for variety in our eating. Daily helpings of the same food from each group may contribute to a balanced diet but not necessarily a healthy one. For example, the vitamin C content of fruits may range from about 4 mg per 100 g in raw pears to 200 mg per 100 g in blackcurrants. Also, eating the same foods all the time is boring.

If someone eats more food than is needed for his or her energy needs, putting on weight is the result! Magazines, newspapers, television, films and advertisements bombard us with body images. They shape our ideas of who is fat, thin or just right. A single number called the body mass index is another more accurate measure of fatness and thinness.

$$\text{body mass index} = \frac{\text{body mass (kg)}}{(\text{body height})^2\ (\text{m}^2)}$$

A body mass index of:

- 20–25 reduces the risk of weight-related health problems
- 25–30 is not likely to have much effect on your health, but don't get any heavier…
- more than 30 means that your heart could suffer if you don't lose weight.

People with a body mass index of 30 or more are said to be **obese**. Their life expectancy (the number of years a person can expect to live) may be shorter than people with a lower index. Also some diseases are more common in obese people.

If someone is obese it is not a good idea to lose too much weight too quickly. Weight should be lost gradually.

B Obesity increases the risk of joint problems and heart disease.

A

P

3 How is body mass index calculated?

4 Briefly explain how values of body mass index may be used to estimate people's risk of developing weight-related health problems.

5 Write a couple of paragraphs for someone who is overweight explaining the importance of having a balanced diet and the problems of obesity.

Summary Exercise

Higher Questions

Extension Questions

5. Feeding the world

By the end of these two pages you should be able to:

- describe and evaluate the potential of biotechnology in relation to world food shortage.

About 6.5 billion people live in the world. To stay alive and healthy all of us need food – about 30 tonnes of it during a lifetime of 70 years. New technologies help farmers work the land so that food is produced in great quantities.

For example, plant **breeding** programmes produce new varieties of high-yielding crops. The new varieties are produced by crossing two different varieties (hybridisation). Each variety has desirable characteristics (e.g. resistance to disease, high yield), and crossing concentrates the desirable characteristics into the new hybrid variety which is then reproduced asexually. For example breeding new varieties of rice plants has boosted yields by more than 25% since the 1970s.

A More people in the world eat rice than any other single food. High-yielding varieties of rice plant help to meet the demand.

P

1 How are plants bred to increase desirable characteristics?

2 What is a hybrid?

3 Give examples of some desirable characteristics bred into plants.

Usually a traditional plant breeding programme takes up to 10 years to produce a new crop variety. Programmes which use **genetic engineering** are much faster. This means that crops with desirable characteristics can be produced more quickly. As a result more food (and better quality food) is available for people – particularly people in developing countries. For example a new variety of rice which makes vitamin A and is rich in iron has been produced because of genetic engineering. Hundreds of millions of people could benefit from this development.

V

Genetic engineering is one example of biotechnology we can use to produce more food. Another example is growing microorganisms for food in huge containers called fermenters. Single-cell protein (SCP) is a food produced from microorganisms. SCP could be useful as a foodstuff because it is fast growing and has a high protein, vitamin and mineral content. Microorganisms double their mass within hours, whereas the plants and animals we eat may take weeks and months to grow to full size.

B Fermenters.

C The bacterium *Spirulina* is a type of SCP that grows in shallow water. The Aztecs made cakes from *Spirulina* for hundreds of years.

The microorganisms multiply using the nutrients in the fermenter as a source of energy and materials for their growth. Using the nutrients is an example of fermentation in action. Nutrient is replaced as it is used up. The fermenters are run for months at a time during which microbial mass is removed at regular intervals and processed into food.

Have you ever wondered?

Can't we already feed the world?

SCP can be grown in large quantities in fermenters which are a controlled environment independent of the weather and which take up little space. The technology is established and can be used in developing countries where food may be in short supply. However, do you think we might be reluctant to eat food made from microorganisms? How can we make SCP more acceptable? The issues which come from our ability to develop GM crops and produce SCP food stir up fierce emotions and vigorous debate. We can do the science.... but should we do it? And if so, how best to do it?

4 What is genetic engineering?

5 Briefly describe the advantages of eating food made from single-cell protein.

6 What are some of the advantages of producing single-cell protein compared with growing crops? State *one* disadvantage.

7 Do you think that biotechnology should be used to increase food production? Explain why. Mention high-yield rice crops, GM crops and SCP in your answer.

Summary Exercise

Higher Questions

Extension Questions

6. Genetically modified plants

By the end of these two pages you should be able to:

- describe the genetic modification of crops to ensure that they are resistant to herbicides
- explain the use of these crops in weed control to reduce loss of food
- describe the use of *Agrobacterium tumefaciens* as a vector to transfer genes coding for herbicide resistance into the genome of a plant cell
- describe the use of the toxin produced by *Bacillus thuringiensis*
- describe how the gene coding for toxin production can be inserted into plants.

A Poppies are weeds that compete with the wheat.

Weeds are plants that grow where we don't want them. Farmers in particular don't want weeds growing among their crops. The weeds compete with the crops for the resources they both need. Farmers use chemicals called **herbicides** to kill weeds. However, herbicides can kill crops as well, so farmers must use the chemicals carefully.

The cells of some plants contain a **gene** which protects plants from the effects of herbicides. The plants are herbicide **resistant**. We can use genetic engineering to introduce into crop plants the gene which controls herbicide resistance. Farmers can then destroy competing weeds with herbicides without harming the crop, making it easier for them to produce more food.

Agrobacterium tumefaciens is a bacterium which lives in the soil. In the early 1980s scientists learnt how to transfer the gene controlling herbicide resistance into crop plants using *Agrobacterium*. They had carried out **genetic modification**.

A plant infected with *Agrobacterium* produces a cancerous growth (tumour) called a crown gall. We can use *Agrobacterium* as a **vector** to transfer genes controlling herbicide resistance into the genetic material of plant cells. Diagram B shows how it is done.

The *Agrobacterium* cell has a small loop of DNA called the *Ti* plasmid (*Ti* stands for tumour inducing). The herbicide resistance gene is inserted into the *Ti* plasmid. The plasmid causes a plant infected with *Agrobacterium* to produce a crown gall. The cells of the gall each contain a *Ti* plasmid with the herbicide resistance gene in place. Pieces of tissue cut from the crown gall are cultured and grow into plants. Each plant is genetically modified to be resistant to herbicide.

1 What is a herbicide?

2 Briefly describe how we can use genetic engineering to produce herbicide-resistant crop plants.

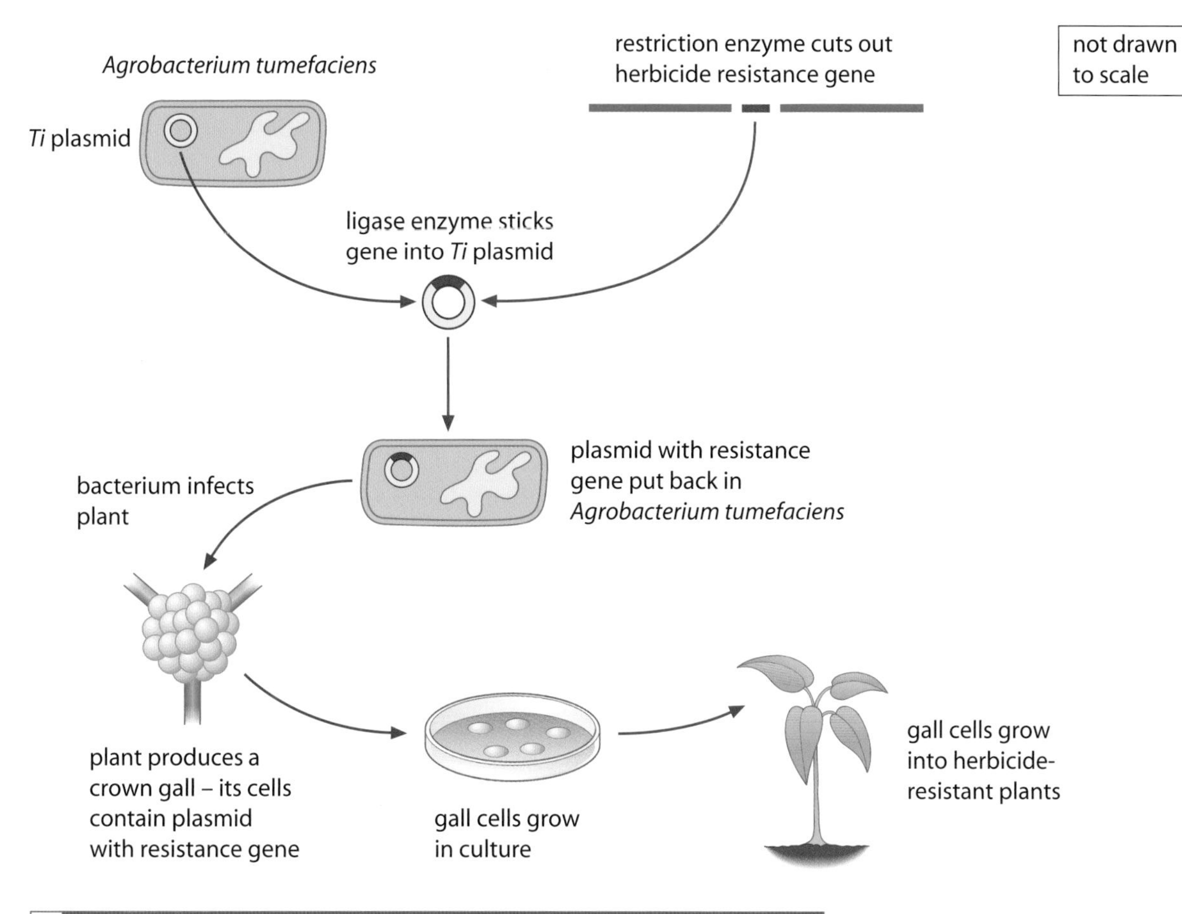

B Producing herbicide-resistant plants using *Agrobacterium tumefaciens.*

Worldwide, insects may eat or damage more than 30% of the crops grown each year. Different insecticides (substances which kill insects) are sprayed onto the crops to protect them from insect attack. However, insecticides are costly and harmful to wildlife and humans.

The bacterium *Bacillus thuringiensis* kills leaf-eating caterpillars and the larvae (young stages) of flies and mosquitoes. The damage is caused by a **toxin** called insecticidal crystal protein (ICP) which is produced by the bacterium. ICP attacks the caterpillar's gut. The caterpillar stops feeding and dies.

Crop plants have been genetically modified using the gene for ICP from *Bacillus thuringiensis* with *Agrobacterium* as a vector. The gene controlling the production of ICP is inserted into the *Ti* plasmid of *Agrobacterium*. Crop plants are infected with the genetically modified bacterium. Pieces of tissue cut from the crown galls which develop are cultured and grown into plants. The plants are genetically modified to produce ICP and are able to resist insect attack.

3 Insecticides are substances which kill insects. State two disadvantages of using insecticides to protect crops from insect attack.

4 What is insecticidal crystal protein?

5 Draw an annotated diagram to show the production of insect-resistant plants.

Summary Exercise Higher Questions Extension Questions

7. Should GM crops worry us?

By the end of these two pages you should be able to:

- discuss the ethics of genetic modification, particularly its use in developing countries.

Growing genetically modified (GM) crops in the UK is a controversial issue. However, new approaches to producing more food are needed, especially in developing countries where climate and soil can make growing crops difficult and food is sometimes in short supply.

Think about the questions:
- how can we choose between what should be done scientifically and what should not be done – what are the **ethics** of genetic modification?
- are the ethics the same for all countries, or are issues which affect developed countries different from those for developing countries?

1 Why do we need new approaches to producing more food?

2 How might industrial goals be different from scientific goals?

Producing GM crops is not just a science. It is also an industry, made possible by science. However, industrial goals are different from those of scientific progress. Industrial goals include making new markets and selling the most product.

Our views are influenced by how we think these goals work for our good, in this case new ways of producing more food. If we are uncertain about the science that makes the goals possible, then we might decide against these new ways. Once doubts set in, then no matter what scientists, business people or politicians say, public opinion is slow to change.

Have you ever wondered?

Is genetically modified food safe to eat?

Many people in the UK object strongly to GM crops and the food which comes from it. Some worries are that:
- it's not natural
- eating GM food may affect our health
- GM crops may harm wildlife
- pollen from crops genetically modified to resist herbicides may transfer to wild plants. If these plants are weeds, there is a danger of weeds developing which are herbicide resistant.

A Many people object to GM crops.

Have you ever wondered?

Do genetically modified organisms harm the environment?

So, what are the benefits of GM crops?

- It is easier and safer for farmers to control the weeds and insects that reduce the amount of food produced.
- The volume of herbicides and insecticides is reduced.
- Reducing the use of herbicides and insecticides reduces risks to wildlife and damage to the environment.

B Growing GM crops.

These advantages are true not only for the UK but for most parts of the world where modern (intensive) methods are used to produce food.

As well as crops genetically modified to resist herbicides and insect attack, other developments include crops modified to:

- grow in places where rainfall is low
- resist the microorganisms which cause crop diseases
- produce their own fertiliser.

The result is improved food production, thanks to GM! The impact on supplies of food in dry, hot, developing countries could be immense.

Have you ever wondered?

Should we be making developing countries buy new seeds every year?

However the issues stirred up by the debate in the UK and other developed countries are the same worldwide. In countries short of food, where does the balance lie between improving food production to save lives and the controversy that surrounds GM technology? Perhaps the answer lies with the public, scientists and industry working together to develop new gene technologies which are acceptable and which will help feed the world's ever-growing population.

3 List some of the worries about growing GM crops.

4 Briefly explain one benefit of growing GM crops.

5 Could the development of GM crops solve the problem of food supply in countries short of food? Explain your answer.

Summary Exercise | Higher Questions | Extension Questions

8. Genetically engineered insulin

By the end of these two pages you should be able to:

- describe the production of insulin using genetic engineering.

Date: 14th October 1980. *Place:* New York Stock Exchange. The launch of a small biotechnology company called Genentech sparked frantic business. Its share price rocketed from US$35 to US$88 within the first 20 minutes of trading. At the end of the day each Genentech share was worth US$71.25. Why was there so much interest in a four-year old Californian company specialising in genetic engineering?

Two years previously, scientists at Genentech had isolated the human genes which carry the code that enables particular cells of the pancreas to produce the hormone **insulin**. They had learnt how to transfer the genes from these cells into bacterial cells. The bacterial cells had been genetically engineered to contain human insulin genes. The genes continued to control the production of insulin even though they had been moved from one type of cell (pancreatic cells) to another type of cell (bacterial cells).

When bacterial cells multiply, the genes which have been transferred into them also multiply. Bacterial cells multiply very rapidly and large numbers of cells soon build up. In the case of bacterial cells genetically engineered to contain the human insulin genes, this means large amounts of insulin can be made quickly.

A Trading at the New York Stock Exchange.

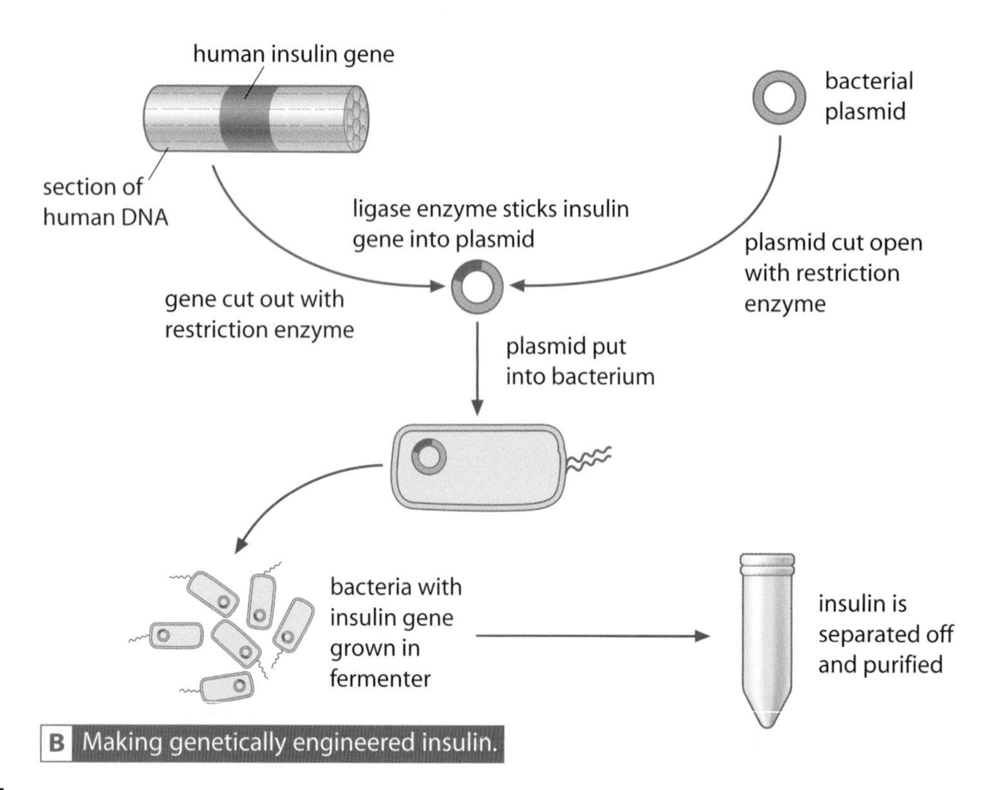

B Making genetically engineered insulin.

The conditions inside huge fermenters encourage the rapid multiplication of bacteria. The fermenters are filled with a solution of all of the substances the bacterial cells need to grow. In the case of bacterial cells genetically engineered to produce insulin genes, the insulin the cells make is released into the solution. It is separated from the solution, purified and packaged ready for use.

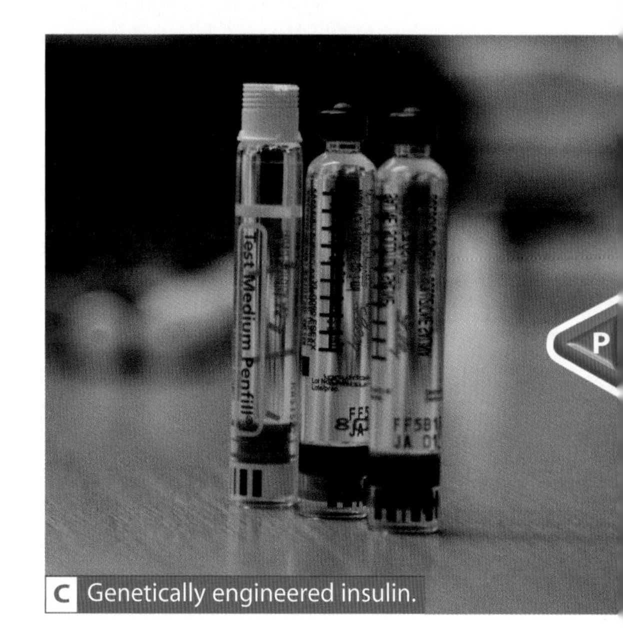

C Genetically engineered insulin.

1 Which tissue in the body produces the hormone insulin?

2 What is a fermenter? Briefly explain how it works.

Why should we want to transfer insulin genes from cells of the human pancreas to bacterial cells? Insulin controls the level of sugar in our blood. Some people do not produce enough insulin and suffer from diabetes. Daily injections of insulin control the condition. Before genetic engineering, people suffering from diabetes depended on insulin sourced from cows and pigs. This was unacceptable to vegetarians and people with certain religious beliefs, it was expensive, and sometimes caused allergies. All of these problems were overcome when genetically engineered insulin became available to treat diabetes. It is cheap and safe to use because chemically it is the same as the insulin our cells produce naturally.

By 1982 genetically engineered insulin was in general use. Genentech's achievement opened the way for the large-scale production of medicines using the techniques of genetic engineering. The medicines are reliable, relatively cheap and, in the case of insulin, more suitable for the human patient.

3 What are some of the advantages of using bacteria which contain human insulin genes to produce insulin?

4 Why is it safer for people who are diabetic to use insulin produced by genetically engineered bacteria than insulin obtained from cows and pigs?

5 Use knowledge from this and previous pages to produce your own diagrams showing the production of genetically engineered insulin. Include the following words: restriction enzyme, plasmid, ligase.

Summary Exercise Higher Questions Extension Questions

9. Genomics

By the end of these two pages you should know:

- how to describe genomics.

DNA consists of thousands of building blocks called nucleotides. Each nucleotide has a base as part of its structure. There are four different bases: adenine (A), thymine (T), guanine (G) and cytosine (C).

A gene is a section of DNA. The differences between one molecule of DNA and another – and therefore one gene and another – are the result of a difference in the order of bases along it.

The order of the bases of our genes is our genetic code. The code is a set of instructions for combining amino acid units in the correct order to make a molecule of protein or part of a protein.

The instructions needed to arrange an amino acid unit in its correct place in a protein molecule are contained in a row of three bases – for example, GCC–ATG–GAT–CAA. The order of the groups of three bases controls the order in which amino acid units are combined, producing a particular protein.

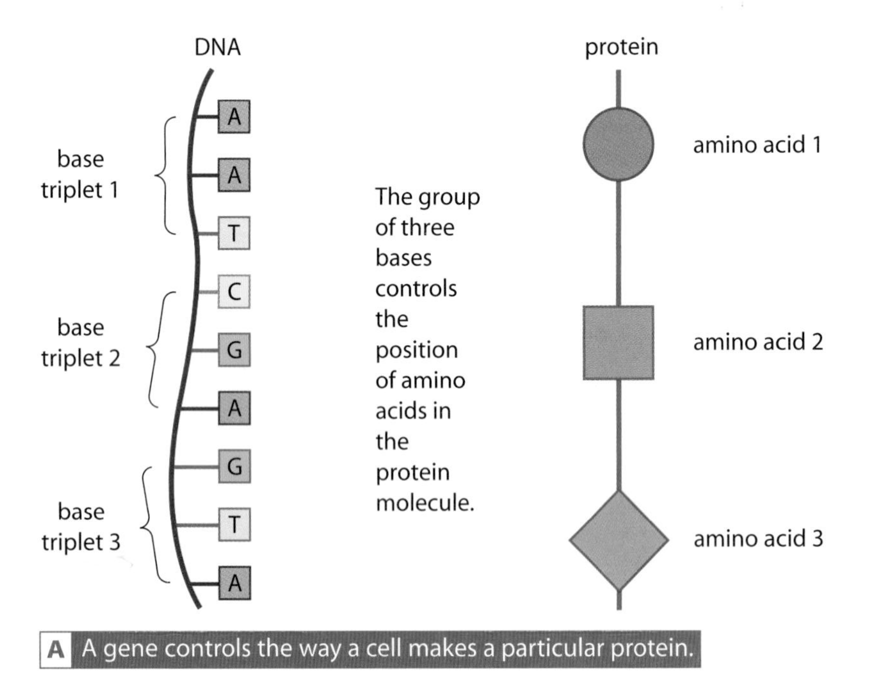

A A gene controls the way a cell makes a particular protein.

A **genome** is all of the DNA in each cell of an organism. **Genomics** is the science of working out the sequence (the order) of the bases in the strands of DNA which make up a genome. In 1986, scientists in the UK and the USA set up the Human Genome Project to sequence human DNA. The scientists have now worked out the order of the bases of the 3.1 billion nucleotides which make up the DNA of the human genome.

1 What is a gene? What does it do?

2 What does the word 'genome' mean?

How did we read the letter sequence of the human genome? Samples of cells were taken from anonymous volunteers. The chromosomes of the cells were broken up into pieces to get at their DNA. Thousands of copies of the pieces of DNA were made, to provide enough material to work on. The DNA was then placed inside machines called sequencers, and the most probable order of bases displayed on a computer screen.

A gene is identified by comparing the sequence of its bases with the sequence of bases in known genes. However, the question is: which gene(s) code(s) for which protein(s)? The new and developing science of bioinformatics is helping to answer this question. Its methods involve the use of powerful computers which help match the base sequences of genes with the proteins they code for. The aim is to:

- find which gene(s) code(s) for which protein(s)
- show in which cells a gene is active (controlling production of its protein)
- show when each gene is active.

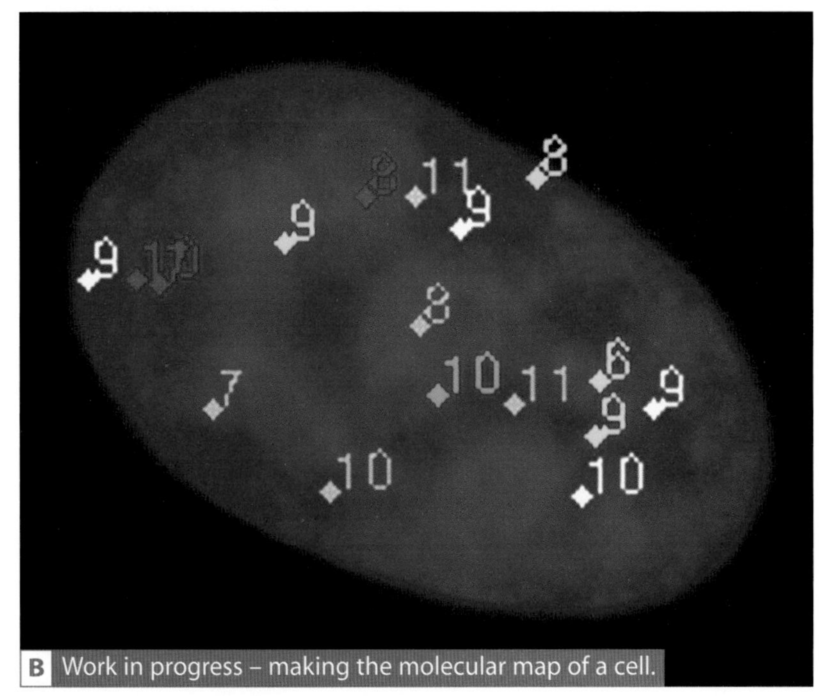

B Work in progress – making the molecular map of a cell.

Bioinformatics is helping to make a molecular atlas of the cell. The map will show where the proteins coded for by our genes are normally found.

3 What is bioinformatics?

4 Summarise the aims of producing a molecular atlas of the cell.

5 Briefly describe the work of the genome project.

C A draft of the human genome was published on 15 February 2001 in the journal *Nature*. This is *The Eagle* in Cambridge, where James Watson and Francis Crick announced in February 1953 that they had discovered the structure of DNA.

P

Summary Exercise

Higher Questions

Extension Questions

10. Genomic medicine

By the end of these two pages you should be able to:

- describe the role of biotechnology in developing new medicines – pharmacogenomics.

Your response to a drug often depends on your DNA. Except for identical twins, each person is genetically slightly different from everyone else. These differences explain why some people benefit from taking a particular drug whereas others receive no benefit and may even be badly affected by it. Millions of people worldwide suffer from these adverse drug reactions and some die because of them. The different responses of people to particular drugs occur because drug companies are limited to developing a single drug to treat a disease. Combining genomics with pharmacology (the science of drugs and their actions on the body) offers the potential to make drug treatments that suit the genetic make-up of each individual. We use the term pharmacogenomics to refer to this new area of science.

1 What is an 'adverse drug reaction'?

2 What is pharmacogenomics?

Today, the use of pharmacogenomics is limited but developments look promising. For example, salbutamol is used to treat people suffering from asthma. The drug acts on the muscles which control the opening and closing of the tubes which take air into and out of the lungs. There are different versions of the muscles. One version is unresponsive to salbutamol. For people with unresponsive muscles, another form of asthma treatment is needed. Analysing the genes responsible for the different versions of the muscles means that people unresponsive to salbutamol can quickly be identified and provided with alternative drugs.

Have you ever wondered?

Will scientists be able to make me a personalised medicine?

Enzymes in the liver are another example of where we might use pharmacogenomics to avoid adverse drug reactions. The enzymes break down more than 30 different types of drug. However, variations in the genes which code for the enzymes means that their action in some people is more effective than in others. The people with the less active forms of the enzymes are at greater risk from adverse drug reactions because the drugs are not broken down. Instead they build up in the body. Genomics could be used to find out how effective a patient's liver enzymes are. The information can then be used to guide the choice of drug and its dosage.

A An asthma sufferer inhales drugs that help to keep her airways open.

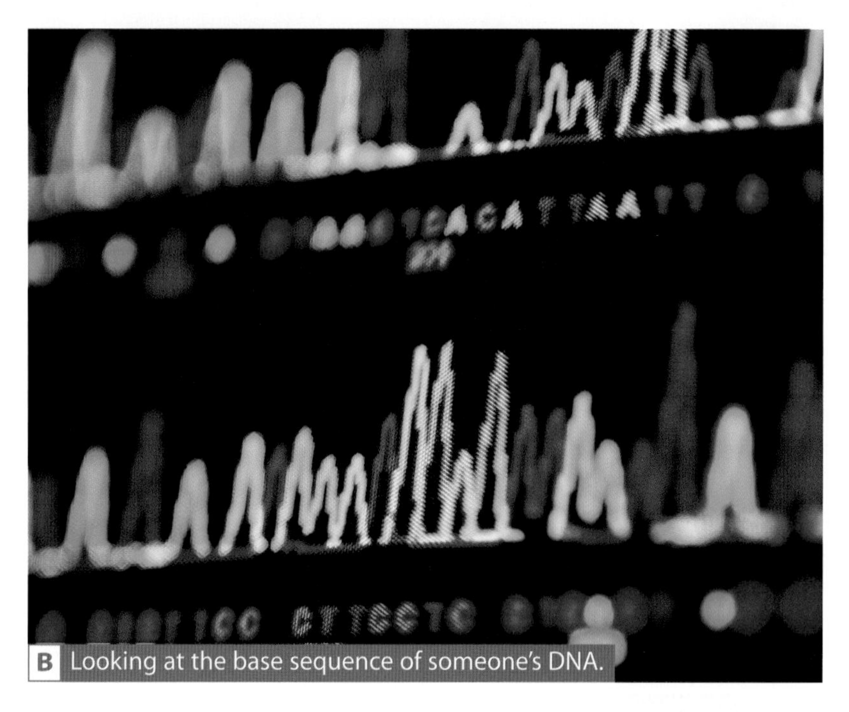

B Looking at the base sequence of someone's DNA.

Eventually it will be possible to personalise drug treatments based on an understanding of our individual genetic variations. These variations decide the effectiveness of drugs and our individual responses to them. This will have the following outcomes.

- *Better medicines.* Drug companies will be able to produce treatments which target particular diseases. Damage to healthy tissues will be reduced.
- *Safer drugs.* The likelihood of adverse drug reactions will be reduced or removed altogether.
- *More accurate drug dosages.* At present drug dosages are based on the age and mass of the person being treated. In the future dosages will be based on the individual's genomic profile – how well the body processes medicines.
- *Better vaccines.* New vaccines will be made from genetic material, and will have the benefits of existing ones without the risk of side-effects. These occur because some of us are allergic to the substances used to make today's vaccines.

Pharmacogenomics promises to take the guesswork out of finding the right drug, delivering the right dose and providing the best treatment for the illnesses which threaten the good health of us all.

3 Why does the response to drugs vary between individuals?

4 Briefly explain the potential importance of pharmacogenomics in the treatment of disease.

Higher Questions

Extension Questions

11. Medicinal plants

By the end of these two pages you should be able to:

- describe the importance and medicinal value of drugs produced by plants.

Long ago our well-being often depended on knowing how the plants we ate affected us. Some plants tasted bitter, some made us sick, and others were poisonous, depending on how much of them we ate. However, we discovered that if we ate less they did not poison us but cured disease and relieved aches and pains. Trial and error taught us that, although too much might mean death, just enough helped us to stay alive. Some of these substances are the basis of many drugs used in modern medicine. Traditional medicine refers to the treatment of ill health using plant-based medicines.

A Traditional medicine at work.

1 Briefly describe the historical importance of traditional medicine in the treatment of disease.

2 Why is traditional medicine important to our modern understanding of how to treat disease?

Today we think of medicine as modern and technologically sophisticated, but many modern drugs still come from plants. Our knowledge of plant-based drugs has usually come from traditional medicine and the reputation of particular plants for curing diseases. This knowledge is the spur for research into new medicines and drugs sourced from plants. For example the *Strychnos* plants of the Amazon rainforest are a source of curare, a substance that relaxes muscles. Hunters in the rainforest tip their blowpipe darts with curare before shooting the birds and monkeys they eat. The muscles which control breathing stop working and the animals suffocate and fall from the trees to the ground. During operations, doctors use low doses of curare to relax the muscles of the patient undergoing surgery, making it easier to control their breathing.

Have you ever wondered?

Who owns the medicine if the original plants come from a different country?

At least 6000 plant species are interesting from a medical point of view. The substances they produce fall into several groups. For example:

- substances which affect the nervous system (e.g. morphine, cocaine and curare)
- substances which control a variety of heart problems (e.g. digitoxin).

These substances can be extracted from plants in the laboratory and made up into medicines and drugs. Studying the chemical structure of the substances allows scientists to synthesise (make) some of them so that we do not have to rely on plants as the source of raw materials. However, the chemical structure of many plant substances is complicated and there is much to learn before the synthesis of many plant-based drugs is possible.

Up to 50% of all the drugs we use to fight disease cannot be synthesised and are still directly sourced from plants. For example, some anti-cancer drugs (e.g. vinblastine and vincristine) are sourced from the rosy periwinkle that grows in tropical rainforests. Treatments using the drugs are highly successful. Almost 75% of the plants known to produce anti-cancer substances grow in tropical rainforests. Yet we know little about the chemistry of the substances themselves. How many more medicinal plant superstars are there waiting to be discovered?

B Scientists continue to search for plants that might be a source of new drugs.

3. Briefly describe the importance of curare in modern surgery.
4. Name a plant-based medicine that affects:
   a the nervous system
   b the heart.
5. What condition can be treated with vinblastine?
6. Describe how plant-based drugs are produced and give some examples.

P

Higher Questions

12. Aspirin and quinine

By the end of these two pages you should be able to:

- recall that salicin is found in the bark and leaves of willow trees and is used for pain relief
- describe the importance and medicinal value of aspirin
- describe the importance and medicinal value of quinine
- recall that quinine comes from the bark of the *Cinchona* tree, and that until the 1930s it was the only real treatment for malaria.

For thousands of years it has been known that the bark and leaves of the willow tree relieve pain and reduce fever. The active ingredient is **salicin** and was extracted from willow bark in 1828. Salicylic acid can be made from salicin. It is a more effective painkiller than salicin but unfortunately it irritates the lining of the stomach and intestines.

A Salicin is found in the bark and leaves of willow trees.

Aspirin contains a version of salicylic acid which reduces pain without such adverse side-effects. Aspirin is sometimes described as a wonder drug, because as well as relieving pain it acts to reduce:

- swelling in joints and other tissues (anti-inflammatory)
- the formation of blood clots
- fever in people whose body temperature is higher than normal.

Aspirin is not a copy of something that comes from nature, although salicin (which can be converted into aspirin) is. It was first synthesised in 1899 and it launched the pharmaceutical industry which, today, is worth billions of pounds worldwide.

1 What does 'active ingredient' mean?

2 What is the active ingredient in the bark and leaves of the willow tree?

3 What is the advantage and disadvantage of using salicylic acid as a painkiller?

4 Why is aspirin sometimes described as a wonder drug?

Malaria flourishes in the warmer tropical regions of the world. The disease is caused by a single-celled parasite called *Plasmodium*. Once the parasites are in the blood stream they attack red blood cells. The infected person suffers flu-like symptoms followed by high fevers, heavy sweating and teeth-chattering chills.

The *Plasmodium* parasite is carried by female *Anopheles* mosquitoes. It passes from person to person when the mosquitoes feed on the blood of people infected with malaria. We say that the female *Anopheles* mosquito is the **vector** of *Plasmodium*.

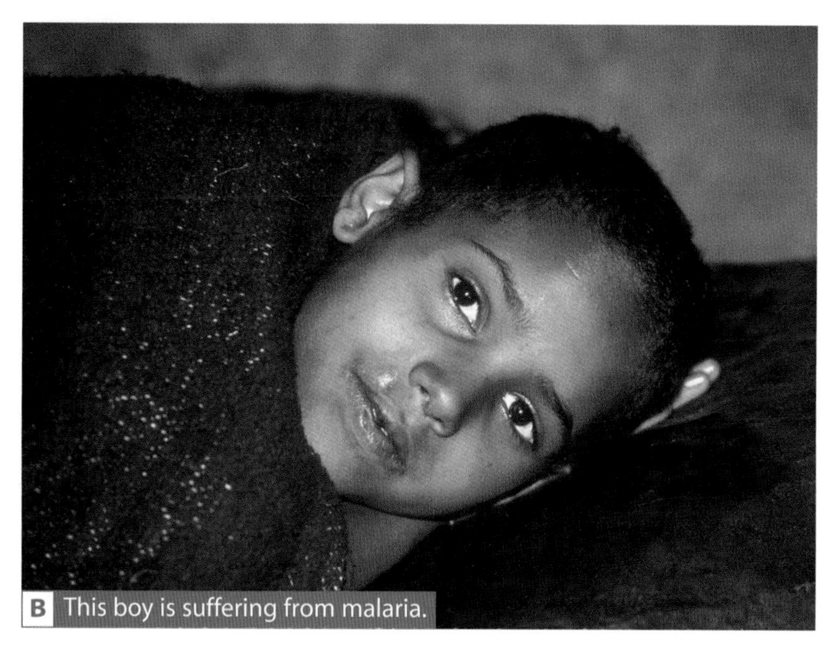

B This boy is suffering from malaria.

C *Cinchona* grows on the slopes of the Andes in Peru, South America.

Since the 1600s we have known that chewing the bark and leaves of the *Cinchona* tree is a useful treatment for malaria. The active ingredient is **quinine** and was extracted from *Cinchona* bark in 1820. The drug lowers body temperature and kills *Plasmodium* parasites in the red blood cells.

For 300 years quinine was the only effective treatment for malaria. In the 1860s plantations of *Cinchona* trees were established in Java and Indonesia from seedlings smuggled out of South America. They were the world's main source of *Cinchona* bark for making quinine until World War II cut off supplies. Antimalarial drugs like chloroquine were developed to fill the gap. They proved to be more effective and easier to make than quinine. However, some strains of *Plasmodium* have become resistant to the drugs, and in some parts of the world quinine is once more the antimalarial drug of choice.

5 Which organism is:
a the malarial parasite
b the malarial vector?

6 What does the word vector mean?

7 Name the first effective drug used to treat malaria.

P

8 Describe how traditional medicine has been used to develop important drugs, using aspirin and quinine as examples.

Summary Exercise

Higher Questions

Extension Questions

13. Artemisinin and taxol

By the end of these two pages you should be able to:

- describe the importance and medicinal value of artemisinin and its derivatives
- recall that artemisinin is extracted from the Chinese plant *Artemisia annua* and is used for treating malaria and reducing its transmission
- describe the importance and medicinal value of taxol
- recall that taxol is derived from yew trees and is used as an anti-cancer agent.

Artemisinin is widely used in China and south-east Asia for the treatment of malaria. It is extracted from leaves of the shrub *Artemisia annua*. Derivatives of artemisinin include artemether and artelinate. Artemisinin and its derivatives kill *Plasmodium*, the parasite that causes malaria. It also prevents *Plasmodium* from reproducing inside its mosquito vector, therefore reducing transmission from person to person.

Unfortunately artemisinin is only active in the body for a few hours. However, combination with other antimalarial drugs increases its effectiveness. This combination treatment is called artemisinin-based combination therapy (ACT) and its use is recommended by the World Health Organization in places where the malaria parasite has developed resistance to the antimalarial drug chloroquine.

The shortage of *Artemisia* leaves has prompted research into ways of making artemisinin in the laboratory. Genetically engineered yeast cells produce a substance like artemisinin which can easily be converted into artemisinin itself. Laboratory trials have shown that the substance is more effective than the natural product, is active in the body for longer and is cheaper to produce.

Artemisinin may also be useful in the treatment of cancer. Early tests show that it damages cancerous cells without affecting healthy ones. It also slows development of the vessels supplying blood to cancerous tissues. The tissue is starved of oxygen and dies.

1 What diseases can the drug artemisinin be used to treat?

2 Why might yeast cells be the key to producing more artemisinin?

A Artemisinin is extracted from the shrub *Artemisia annua*.

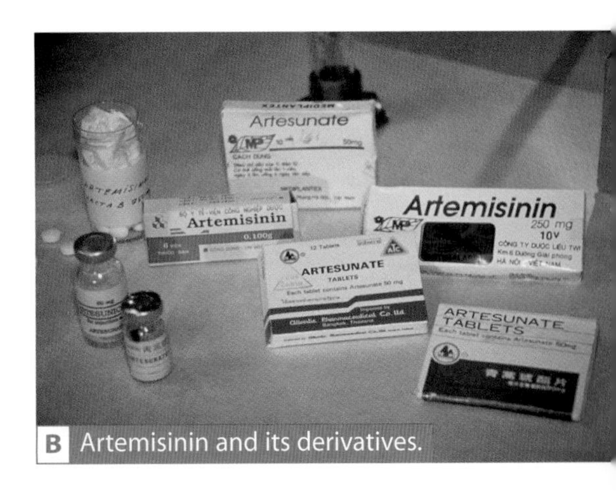

B Artemisinin and its derivatives.

In the early 1960s, the US National Cancer Institute set up a programme to look for new anti-cancer drugs sourced from plant materials. Success came in 1967 when a substance with anti-cancer activity was extracted from the bark of the Pacific yew tree. The substance was **taxol**.

C Taxol is extracted from the Pacific yew tree.

The chemical structure of taxol was worked out in 1971 and by 1980 the effects of taxol on cells were understood. It stops cells from dividing. The cells die. Cancerous cells are affected more than healthy ones.

There was so much demand for taxol as a possible anti-cancer drug that demand for bark from the Pacific yew tree soared. However, the tree is one of the slowest growing in the world and is a protected species. Only a small amount of taxol can be extracted from the bark, and the tree is killed in the process.

Luckily a substance similar to taxol was discovered in the needle-like leaves and twigs of the European yew tree. The substance was chemically modified to form a semi-synthetic version of taxol called paclitaxel. This became available for the treatment of a variety of cancers in 1995.

European yew trees quickly replace their needles. Taking large quantities of needles therefore has little effect on the number of yew trees. However, producing enough paclitaxel to meet demand is costly and limits use of the drug. In the future, scientists hope to be able to synthesise the anti-cancer part of taxol in the laboratory. Alternatively, yew tree cells could be grown as a cell culture on a large scale, producing taxol.

3 Briefly describe the effects of taxol on cells.

4 What is cell culture? How could cell culture be used to make large quantities of taxol?

5 Describe how scientific technology could help to produce drugs more cheaply, using artemisinin or taxol as an example.

P

Summary Exercise

Higher Questions

Extension Questions

14. Stem cell research

By the end of these two pages you should be able to:

- describe stem cell research and therapies
- describe the use of stem cells as possible treatments for diseases such as Parkinson's disease.

Once fertilised, an egg cell soon divides into two by mitosis. Repeated divisions follow and a hollow ball of cells called the embryo develops. The cells on the inside of the embryonic ball are all the same. They are embryonic **stem cells**.

Stem cells are unspecialised. However, as cell division continues and the embryo develops into a foetus, the cells begin to change into the different types of cell that form the tissues and organs of the body. The process is called differentiation. Once differentiation is complete, the foetus grows into a fully formed baby.

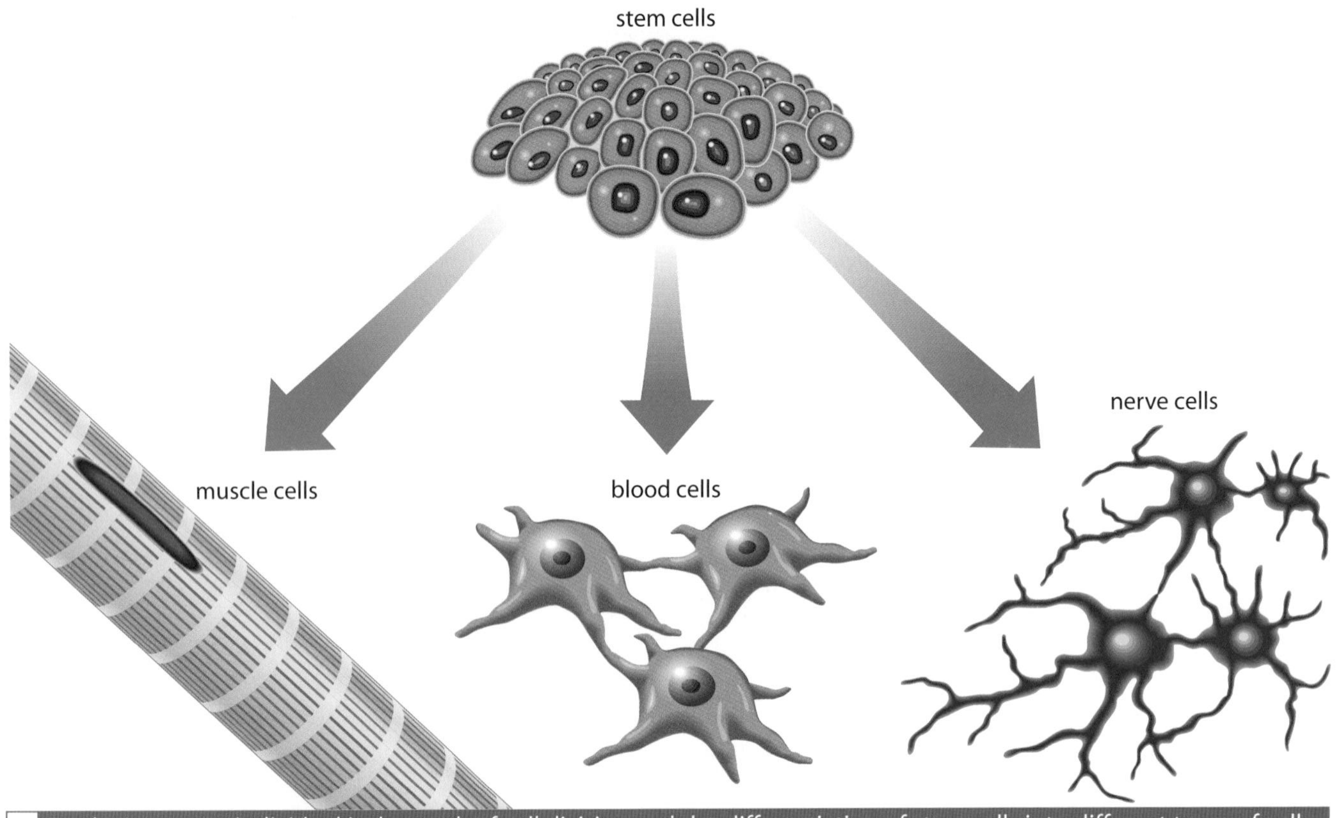

A Producing a new individual is the result of cell division and the differentiation of stem cells into different types of cells.

P

1 What is mitosis?

2 What are stem cells?

3 What is cell differentiation?

Even as adults, some of our stem cells remain. For example, there are adult stem cells in the bone marrow (found in the hollow centre of the leg bones and arm bones). These give rise to blood cells, replacing old, worn-out cells with new ones.

It seems likely that small numbers of stem cells remain in other body tissues as well. If injury or disease damages a tissue then its stem cells divide and differentiate into new cells, repairing the damage. Now we can understand why stem cell research causes so much excitement. If we can encourage embryonic or adult stem cells to multiply and differentiate, unlimited supplies of different types of cells will be available to treat people whose tissues are so damaged as to be beyond self-repair. This is called stem cell therapy.

Stem cells could be the source of new cells to replace diseased and damaged tissues in conditions such as **Parkinson's disease**, diabetes, different cancers and Alzheimer's disease. The work of scientists researching stem cells includes finding and isolating new types of stem cell, understanding how the cells work and finding ways of using them to treat disease.

A small cluster of cells in the brain produce a chemical neurotransmitter called dopamine. People with Parkinson's disease do not produce enough dopamine. Their movements are jerky and uncoordinated because the lack of dopamine affects muscle control.

A Assessing the muscle control of a person with Parkinson's disease.

In 1999 US scientists took 10–15 healthy stem cells from the brain tissue of a person suffering from Parkinson's. These unspecialised cells were grown in a solution of all the substances needed to keep the cells alive, and were encouraged to differentiate into the type of cell which produces dopamine. These cells divided by mitosis into millions of cells which were put back into the person's brain tissue. The person's muscle control improved by 40–50%. Such success shows what may be possible when stem cells are used to treat disease. However, the clinical trials studying the effects of stem cell treatments have only been carried out on small numbers of people so far and there are some adverse side-effects, so the results are still uncertain.

4 Currently, are stem cells used to treat disease?

5 What is stem cell therapy?

6 Briefly explain how using stem cells might be used to treat Parkinson's disease.

P

Summary Exercise

Higher Questions

Extension Questions

15. Girl or boy?

At the end of these two pages you should be able to:

- explain that allowing people to choose the sex of their baby may skew the sex balance of the population
- explain that allowing people to choose the sex of their baby may lead to other choices being permitted, e.g. eye colour.

Photograph A shows the chromosomes that decide a person's sex. The larger of the two is the X chromosome; the smaller is the Y chromosome. Remember that:

- a baby's sex depends on whether a woman's egg is fertilised by a sperm carrying an X chromosome or one carrying a Y chromosome
- the birth of (almost) equal numbers of girls and boys occurs because 50% of a man's sperm have an X chromosome and 50% have a Y chromosome.

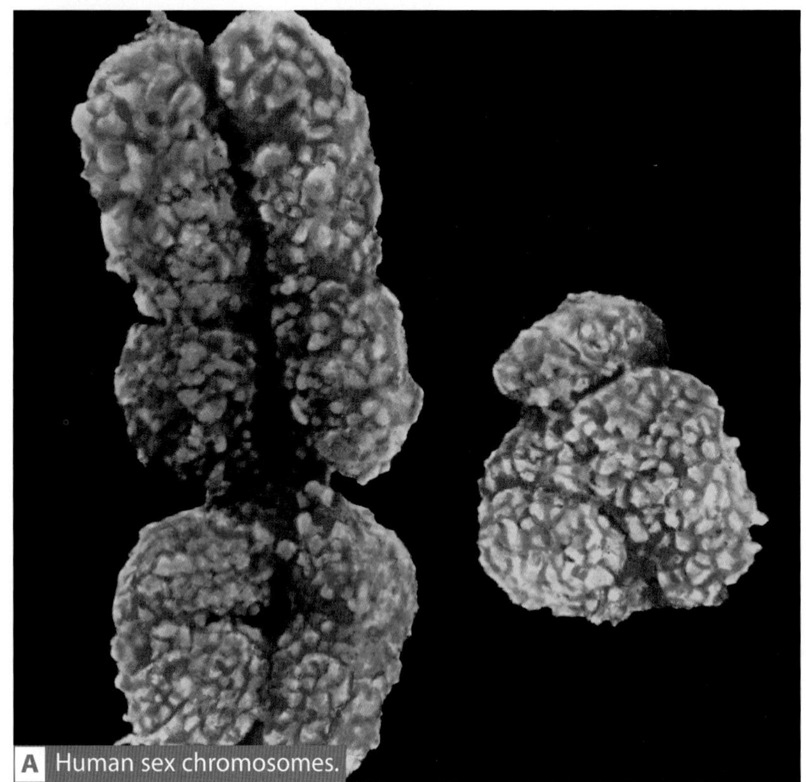

A Human sex chromosomes.

In *in-vitro* fertilisation (IVF), eggs are taken from a woman and sperm are added. Two or three fertilised eggs are chosen for transfer back into the woman's womb. The embryos consist of a ball of cells. Scientists can take one of the cells without harming the embryo, and test it to see if it has genes that could cause disease. Only embryos shown to be free of faulty genes are chosen to be put back into the woman's womb. Finding out the sex of the embryo is also possible.

1 Which sex chromosome(s) are found in:
a a woman's body cell
b a sperm cell
c an egg cell
d a man's body cell?

2 Who determines the baby's sex: mum or dad?

3 What is IVF?

Some genetic disorders like Duchenne muscular dystrophy, fragile X and haemophilia mostly occur in boys. If there is a family history of male-linked genetic disorders, then testing to find out the sex of an embryo can be useful. A female embryo can be chosen instead of a male embryo, as females are less likely to develop the disorders. To select the sex of a baby, doctors can either test an embryonic cell to see which sex chromosomes are present, or the father's sperm can be sorted and only sperm carrying either the X or the Y chromosome are used to fertilise the egg.

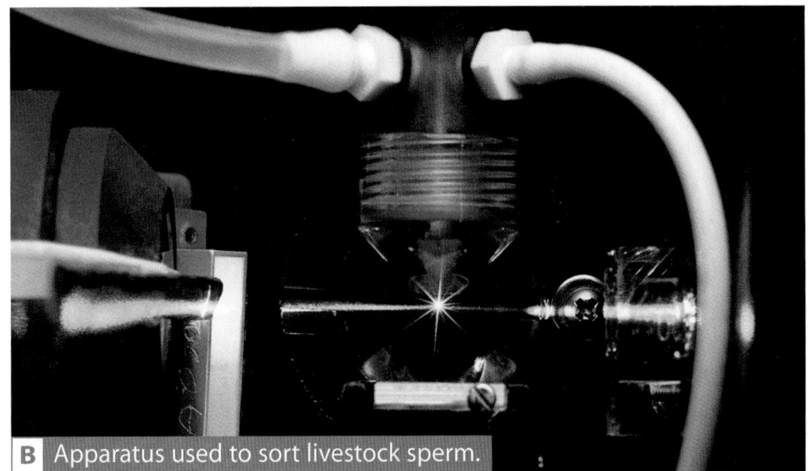

B Apparatus used to sort livestock sperm.

Have you ever wondered?

Should you be allowed to choose the sex of your baby?

In Britain, parents are allowed to choose the sex of a baby to avoid genetic disorders linked to a particular sex, but not just because they want a girl or a boy. If people are allowed to choose the sex of their baby then the 50:50 balance of girls and boys can become skewed. In the case of choosing for reasons of genetic well-being, the balance would probably tip in favour of girls as there are more genetic disorders linked to the presence of a Y chromosome than the X chromosome.

Have you ever wondered?

Are we able to cure genetic diseases?

Testing embryos and choosing the ones with healthy genes, or choosing the sex of embryos so that faulty genes do not show themselves, seems a brilliant way of improving human well-being. However, some people see these developments as unethical. They think that if we are given more freedom to pick and choose sperm, eggs or embryos then it will only be a short step to our choosing the characteristics we want to see in our babies. Choosing the genes which control intelligence and other characteristics would allow us to produce designer babies.

4 Briefly explain two different ways of choosing a baby's sex.

5 How could choosing babies' sex affect the sex balance of the population?

6 What are designer babies?

7 Write a paragraph describing the advantages and disadvantages of allowing people to choose the sex of their baby.

Summary Exercise | Higher Questions | Extension Questions

16. Ethical issues

At the end of these two pages you should be able to:

- describe the ethical implications of reproductive research.

Science enables us to do amazing things. However, sometimes the question arises – should we do them? These are ethical decisions that only we can make.

In 1997 Dolly the sheep was born. She was the first mammal to be cloned from an adult cell. Remember that living things which are genetically identical are called clones. Since Dolly's birth, the ethical issues of cloning animals have been hotly debated.

Today sheep, cattle, pigs, mice and other mammals have been cloned. We are mammals. If it is possible to clone Dolly and other mammals, then it should not be too difficult to clone people. We need to make an ethical decision about whether it would be right or wrong to do this.

A Millie, Christa, Alexis, Carrel and Dotcom are the world's first cloned pigs, born in 2000.

Here are a number of points for you to think about.

- Healthy cells from a sick person can be cloned and used to repair that person's damaged tissues.
- A person needing a transplant could make use of a brain-dead clone of himself/herself as a source of tissues and organs for transplantation. The person's body would not reject the transplanted material because it is genetically identical.

- For some people, the thought of using a brain-dead clone as a source of tissues and organs for transplantation is morally wrong. They argue that respect for human life should outweigh the possible benefits to the patient.
- We can genetically modify cows to produce hormones such as insulin in their milk – we could then produce many of these cows by cloning. Should we genetically modify humans? Would their children be genetically modified?

B GM calves.

In 2003, Dolly was found to have arthritis in her joints. She developed lung disease as well and was 'put down' to prevent her suffering distress. By sheep standards Dolly was young. Did she age more quickly than normal sheep? Did she die young because she was a clone? These questions raise important issues about the safety of cloning and whether we should extend the work to humans – this is currently illegal.

1 Say whether you think the following are ethical or unethical, and give a reason:
a cloning pigs
b cloning humans
c cloning someone's skin and using it to treat their burns
d using a brain-dead person as a source of tissues and organs.

2 Might clones age more quickly than normal? Give a reason for your answer.

Many people think that research into human reproduction could lead to human cloning, designer babies and much else. For example, if a child has a life-threatening disease that could be treated with stem cells, IVF scientists can help the parents to produce an embryo which is a near genetic match to the older child. The embryo is put back into the mother where it develops into a healthy baby. When the baby is born, the life-saving stem cells are taken from the umbilical cord and used to treat the sick older child. The baby is not hurt and no cells are taken from the baby itself. However, the younger child is a designer baby, brought into the world to save someone else.

3 Briefly describe the controversy which surrounds stem cell research.

4 Do you think that scientists developing new reproductive technologies are 'playing God' and creating monsters? Write a paragraph which includes the words cloning, IVF, sex selection and stem cells.

Summary Exercise | Higher Questions | Extension Questions

17. Questions

D

P

1 Plants infected with the bacterium *Agrobacterium tumefaciens* develop a cancerous growth called a crown gall. A loop of DNA called the *Ti* plasmid is responsible. Explain why *Agrobacterium* is an ideal vector for introducing the gene which controls herbicide resistance into plant cells.

2 How does the bacterium *Bacillus thuringiensis* kill insects?

3 What is a balanced diet? Compare a meal of sausage, eggs and chips with a meal of fish, salad, fruit and wholemeal bread. Explain why you might think that the second meal is better balanced than the first meal.

4 Imagine that you want to set up a business using microorganisms to produce foods. In particular you want to produce yoghurt.
- **a** Which food is the starting point for making yoghurt?
- **b** Which microorganism would you use to convert the food into yoghurt?
- **c** Why must hygiene during yoghurt making be strictly controlled?

5 Explain why methods which use genetic engineering to produce new crop varieties are faster than methods which use traditional plant breeding programmes.

6 A newspaper reported the birth of Dolly the sheep with the following headline and article:

Warning on human clones

Fears follow production of sheep from single cell

The chilling prospect of a woman giving birth to an 'identical twin' of her own father was raised by doctors yesterday after the announcement that scientists have for the first time succeeded in creating a clone of an adult animal.

Jeremy Laurance and Michael Hornsby, THE TIMES, *24/02/1997*

- **a** Suggest a reason for the newspaper using such a headline.
- **b** Explain why some people might think cloning of humans is unethical.

Stem cells have been used in the treatment of Parkinson's disease.
- **c** What are stem cells?
- **d** Explain the role of stem cells in this treatment.

7 Genetically engineered yeast cells are used to make vegetarian cheese. The stages are:
- An enzyme is used to cut open a plasmid.
- A gene that controls the way cells make an enzyme important for making cheese is inserted into the plasmid.
- The plasmid is inserted into the yeast cells.
- The yeast cells are put into a fermenter.

- **a** What is a plasmid?
- **b** What type of enzyme is used to cut the plasmid open?
- **c** Explain why the yeast cells are put into a fermenter.
- **d** Name the enzyme which is produced by the genetically engineered yeast cells used to make vegetarian cheese.
- **e** Cheese (vegetarian and non-vegetarian) is made from milk. Explain why non-vegetarian cheese is not acceptable to vegetarians.

8 The *Cinchona* tree is a source of a plant-based medicine.
- **a** Explain why people who live in the Amazon rainforest sometimes chew the bark and leaves of the *Cinchona* tree.
- **b** Name the medicine extracted from the *Cinchona* tree.
- **c** What name describes the combination of genomics and pharmacology in developing new medicines?
- **d** Explain how the treatment of diabetes has been improved using biotechnology.

9 In 1967 a substance with anticancer activity was extracted from the bark of the Pacific yew tree.
- **a** Name the substance.
- **b** Explain the anticancer activity of the substance.
- **c** Why are Pacific yew trees no longer used as a source of the substance?

10 a Explain why a baby's sex depends on the father.

b Some genetic disorders mostly show themselves in boy babies. Explain why this is the case.

c How might genetic testing help deal with these genetic disorders?

d Explain *one* disadvantage that might be caused by this genetic testing.

11 The table shows the results obtained when three different types of yeast were each allowed to ferment a standard sugar solution completely at 16 °C. Six readings for each type of yeast were recorded. Each reading is the time taken, to the nearest hour, from the beginning to the end of the fermentation.

Type of yeast	Time for completion of fermentation of a standard sugar solution at 16 °C (hours)					
A	48	47	41	38	36	42
B	15	30	24	12	16	15
C	24	29	30	32	36	33

a Using the data in the table, calculate the mean time for each type of yeast to complete fermentation. Show your working.

b Suggest which one of the yeasts would be most suitable in bread-making.

c Explain the role of yeast in bread-making.

12 Use your own knowledge and the following information to answer questions **a–d**.

About 2.5 billion people are at risk from malaria, 200 million have severe attacks and about 1.5 million people die of the disease each year. Malaria is caused by the single-celled parasite *Plasmodium*. The parasite is transmitted by the *Anopheles* mosquito.

a How many people die from malaria each year?

b Name the vector of the malaria parasite.

c The drug quinine was the first effective drug used to treat malaria.

i Name the plant from which quinine can be extracted.

ii What are the effects of quinine on a person suffering from malaria?

d Explain the thinking behind using artemisinin-based combination therapy to treat malaria.

18. Glossary

***amino acid** The basic unit from which proteins are made. There are 20 different amino acids.

***artemisinin** A substance extracted from the leaves of the shrub *Artemisia annua*. It is widely used in China and south-east Asia for the treatment of malaria.

***bacteria** Microorganisms that lack a distinct nucleus and many of the other structures found in plant and animal cells.

***biotechnology** The processes where we use plant cells, animal cells and microorganisms to produce useful substances.

***breeding** Reproducing offspring.

***cholesterol** An oily lipid which is an important component of cell membranes. A person with high cholesterol in the blood may be more at risk of developing heart disease than a person with lower levels.

***chymosin** One of two protein-digesting enzymes which form rennet. When added to milk, chymosin causes the formation of solid curds.

***citric acid** A weak acid found mainly in citrus fruit.

***enzyme** Protein which catalyses the chemical reactions which digest food in the gut and the thousands of chemical reactions which take place in cells.

***ester** Carbon compound made from the reaction of an organic acid with an alcohol.

***ethics** The study of values and customs of a person or group – includes concepts such as right and wrong, and responsibility.

***fermentation** When microorganisms break down large molecules in the absence of oxygen (anaerobic conditions) to produce different substances including foodstuffs and drugs. Some of the reactions may be aerobic (oxygen present) but are still referred to as fermentation.

***filtration** The process where a liquid is passed through a fine mesh to remove small particles.

***gelling agent** A substance added to food to thicken it and make it less runny.

***gene** A section of DNA which controls the synthesis of a protein or part of a protein.

***genetic engineering** The techniques used to identify, isolate and insert useful genes from the genetic material of cells of one species into the genetic material of the cells of another species.

***genetic modification** A species into which genes are transferred is said to be genetically modified (GM). The genes are transferred using the techniques of genetic engineering.

***genome** The sequence of bases of all of the genetic material of an organism.

***genomics** The science of working out the sequence (order) of the bases of the molecules of DNA which make up a genome.

***glutamic acid** An amino acid.

***herbicide** A chemical used to kill weeds.

***insulin** The hormone which lowers the level of glucose in the blood.

***invertase** An enzyme (also called sucrase) which catalyses conversion of the sugar sucrose into glucose and fructose.

***lactic acid** Produced when lactic acid bacteria ferment the sugar lactose in the absence of oxygen (anaerobic conditions).

***lactose** A complex sugar made from a combination of the simple sugars glucose and galactose.

***malaria** A disease characterised by a regular cycle of high fever and chills, caused by the parasite *Plasmodium*.

***microorganism** An organism which can only be seen under a microscope. Sometimes called a microbe.

***obesity** A condition where a person's body mass index is 30 or more (with exceptions, e.g. athletes, pregnant women).

***oligosaccharide** A carbohydrate molecule formed from 3 to 10 simple sugar units joined together.

***Parkinson's disease** A disease characterised by the patient's inability to control muscle movements due to a lack of the neurotransmitter dopamine.

***pasteurisation** The process of heating food to reduce the numbers of harmful microorganisms.

***prebiotics** Foods which contain oligosaccharides. The sugars are food for beneficial bacteria in the gut, promoting increase in their numbers.

***quinine** An antimalarial drug originally extracted from the bark and leaves of the *Cinchona* tree.

***resistance** The process where disease-causing organisms can withstand the action of drugs which previously would have killed them.

***salicin** A substance originally extracted from the bark of willow trees. It relieves pain and reduces fever. It can be converted into salicylic acid, which is more effective and is the basis of aspirin.

***stem cells** Cells which can develop and change into the different types of cell by differentiation. Each body tissue contains clusters of stem cells. These are adult stem cells. The cells of the early embryo are undifferentiated and are called embryonic stem cells.

***taxol** An anti-cancer drug originally extracted from the bark of the Pacific yew tree.

***toxin** A poison.

***vector** A carrier of genes from the cells of one species to the cells of another species; a carrier of disease-causing organisms from one individual to another individual.

***yeast** A single-celled fungus.

*glossary words from the specification

Behaviour in humans and other animals

A Parental behaviour and conditioning in birds.

Behaviour is the word used to describe the reactions of animals to what is going on around them. Their behaviour is controlled by the nervous system and hormones. It may be instinctive or learned. Instinctive behaviour is automatic and follows a fixed pattern. Learned behaviour is the result of experience.

Types of behaviour include communication, feeding, courtship and parental care. They can be studied in humans and other animals using techniques from biology and psychology.

In this topic you will learn that:

- animals have evolved instinctive behaviours, through natural selection, which increase their chances of survival
- animals learn throughout their lives to increase their chances of survival and reproduction
- feeding behaviours maximise animals' chances of finding sufficient food
- reproductive behaviours maximise animals' chances of successfully passing on their genes
- social behaviours and communication skills enable animals to respond in particular ways to members of their own species and to members of other species
- humans have made use of other animals in different ways, and there is an increasing awareness of animal welfare issues that need to take account of animal behaviour.

Look at these statements and sort them into the following categories:

I agree, I disagree, I want to find out more

- Behaviour is inherited from your parents.
- Only humans use language to communicate with one another.
- Courting behaviour comes before sexual reproduction.
- Society developed because farming techniques developed.
- The rights of chimpanzees are as important as human rights.
- Testing the safety and usefulness of new medicines on animals is essential before the medicines are used to treat people.

1. Instinctive behaviour

By the end of these two pages you should be able to:

- explain that animals inherit certain patterns of behaviour from their parents, known as instinctive behaviour.

A person steps on a drawing pin, jerks his foot away and shouts in pain. The pin's sharp point pricking his skin is the stimulus for his **behaviour**. He does not think about jerking his foot away from the pin. His action is an example of a reflex response which is automatic, takes a split-second and is **instinctive**.

Instinctive responses are **inherited** from parents and reinforced by **natural selection**. The quicker individuals respond, the better their chances of survival. For each particular stimulus triggering an instinctive response, the response is the same. For example, tickling a baby's foot causes the baby to instinctively fan its toes.

1 What is a reflex response?

2 Explain the statement: 'instinctive behaviour is inherited.'

Woodlice live under stones and in other damp places. In dry conditions they lose water and can die. How do woodlice find damp living space? The response of woodlice (their walking activity) depends on the stimulus (how dry the air is). In dry conditions, woodlice that are active have more chances of finding damp places. When they find somewhere damp they become less active.

Woodlice also instinctively respond to the intensity of light. The lighter it is, the more active woodlice are. This increases their chances of survival because high light intensity usually means lack of shelter and therefore drier air.

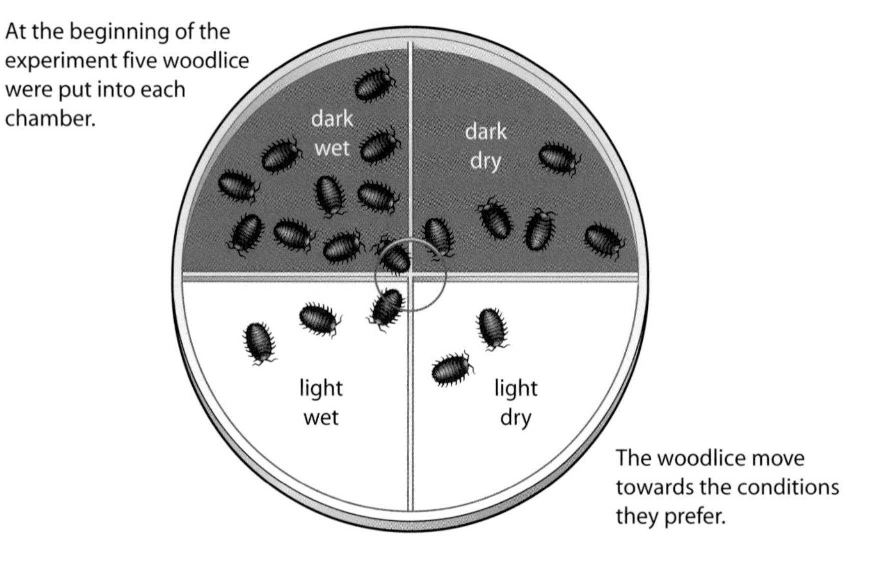

A In a choice chamber, woodlice move towards the conditions they prefer.

Soon after it hatches, a herring gull chick begins to peck at its parent's beak. The chick wants food. The parent soon brings back up into its mouth half-digested food, which the chick eats. The behaviour is instinctive. It allows the chick to respond to a particular stimulus (the sight of its parent's beak) while ignoring others. In this case the chick's response to the particular stimulus of its parent's beak results in the food it needs.

B The herring gull chick pecks at its parent's beak to get a meal.

If you look closely at photograph B you can see a red spot on the adult's yellow beak. The chick pecks at the spot more often than the rest of the beak. The spot is a **signal** to the chick. If the spot is painted over yellow, the chick pecks the adult's beak less often.

In an experiment, newly hatched herring gull chicks were shown wooden models of an adult gull head. The yellow beak of each model had a spot of a particular colour. The beak of one of the model heads did not have a coloured spot. The model with the red spotted beak was pecked by the chicks three times more often than the model with no spot on the beak. The models which had spots on their beaks in other colours were pecked more often than the model with no spot, but less often than the model with the red-spotted beak. The results show that herring gull chicks respond best to the colour red on a yellow background. The red spot is the trigger for the instinctive behaviour which makes it more likely that a herring gull chick will get a meal.

A

3 Explain how the behaviour of woodlice enables them to find damp places to live.

4 What happens when a herring gull chick pecks at the red spot on its parent's beak?

P

5 Describe what instinctive behaviour is, and give examples of instinctive behaviour shown by:
a babies
b adults.

Summary Exercise

Higher Questions

Extension Questions

2. Learned behaviour: conditioning

By the end of these two pages you should be able to:

- explain that an animal's early experiences in life have a big impact on the way in which it behaves as an adult
- explain that animals can learn through conditioning
- recall that humans can make use of conditioning when training captive animals for specific purposes.

A Studying behaviour in crows.

Behaviour which changes in the light of **experience** is **learned**. Pets will show you learned behaviour in action. For example, heading for the kitchen cupboard will soon grab the attention of your cat if that is where you keep its food. The possibility of food links the responses of the cat to your movements.

Have you ever wondered?

What instincts are you born with and what do you learn?

To begin with, a young pet cat will only show interest when the food is seen and tasted. In other words, seeing and tasting are primary stimuli which the cat associates with food. With age and experience, it learns to associate other stimuli, called secondary stimuli, with food – for example its owner heading for the food cupboard. The secondary stimulus itself is not directly linked to the possibility of food, but the cat has learnt the association. We say that the cat has become **conditioned**. The Russian Ivan Pavlov was the first to study conditioned responses scientifically. He investigated the production of saliva by dogs in response to food and other non-food stimuli.

B Ivan Pavlov (1849–1936) – white-bearded, and sitting near to one of the dogs he used to study conditioning.

1 What is the difference between instinctive behaviour and learned behaviour?

2 How do you condition an animal's behaviour?

Pavlov noticed that when food (the primary stimulus) was placed in a dog's mouth the flow of saliva increased. He also noticed that the flow of saliva increased as soon as the dog smelt his hand (a secondary stimulus) – even before the food was placed in its mouth. The dog's production of saliva was increased when Pavlov's personal smell was followed with the taste of food. After a period of presenting the dog with both his personal smell and the taste of food, Pavlov found that his personal smell alone was enough to make the dog produce as much saliva as if it had been given food.

Pavlov also conditioned dogs to produce saliva in response to other stimuli, such as the ringing of a bell. This type of conditioning is called classical conditioning. It fades unless it is reinforced from time to time.

Trial-and-error learning is another type of conditioning. This time the learning develops because of reward or punishment. The American scientist B. F. Skinner (1904–1990) set out to investigate trial-and-error learning in rats and other animals.

C A Skinner box: the rat presses the lever to receive food.

The lever in a Skinner box is the key to an animal's learning. If pressing the lever means it will get food, the animal will quickly associate pressing the lever with reward and keep on pressing. If pressing the lever means an unpleasant stimulus (a mild electric shock for instance), the animal will quickly associate pressing the lever with punishment and leave the lever alone.

You can probably recognise when some of your own behaviour is the result of this type of conditioning. Early experiences of reward and punishment affect our behaviour as adults. Both types of conditioning are used to train animals. Pets are conditioned to respond to their owners' commands. Other animals are trained to do work, e.g. in circuses, in films, as police sniffer dogs, guide dogs.

3 What is trial-and-error learning?

4 Draw a flowchart to summarise the stages in Pavlov's experiment.

5 Explain the differences between classical conditioning and trial-and-error learning.

Summary Exercise

Extension Questions

3. Imprinting and habituation

By the end of these two pages you should be able to:

- explain that an animal's early experiences in life have a big impact on the way in which it behaves as an adult
- explain that habituation is an important part of the learning process in young animals.

Why would a newly hatched duckling follow the cat as if it were one of its parents? It seems a strange scenario. However, young **birds** will often follow the first moving object they see after hatching. We say that they **imprint** on the object.

A Imprinting.

Usually the first moving object a young animal sees is one of its parents. Imprinting on the parent allows the young animal to recognise and follow its parents from an early age. Parents mean food, protection from predators and even shelter. Imprinting therefore improves the chances of young animals surviving beyond the early, possibly dangerous, days of their lives.

Imprinting is a form of learning. However, unlike other forms of learning, imprinted behaviour is not easily changed. The idea is often used in training animals. For example, if a dog's owner becomes imprinted in the young animal's mind, the animal becomes attached to its owner and is much easier to train. The attachment is reinforced by food and the owner's company, and becomes permanent.

B The Austrian scientist Konrad Lorenz (1903–1989) with young geese imprinted on him.

Attachment occurs during a sensitive period, early on in an animal's life. Imprinting will not occur outside the sensitive period. This is why it is much easier to train a puppy rather than an older dog. Imprinting not only affects the links between young and parents, but also affects the social behaviour of adults. For example, young geese which have imprinted on ducks or even humans might attempt to court and mate with them when older.

1 How do young animals benefit from imprinting?

2 Explain the statement: 'imprinting is a form of learning.'

You might step into the shower only to step out again – quickly. The water's too hot … but you want your shower! You know the hot water won't harm you, so you try again – and again, until you step in and you're comfortable. The temperature of the water hasn't changed but you've got used to it. Getting used to a stimulus which you avoid to begin with is called **habituation**. For example, if you buy a coat to keep warm, at first the coat might seem heavy on your body. However, you soon get used to it and don't think about its weight any more. Habituation means that you no longer respond to the weight of the garment.

Habituation is a form of learning which enables animals not to waste time and energy responding to stimuli which might be alarming to begin with but are not harmful. For example, birds fly away when they first see a scarecrow standing in the middle of a field of crops. However, they soon learn that the scarecrow is harmless and ignore it when they return to feed – much to the farmer's annoyance.

C Habituation means that birds ignore the scarecrow.

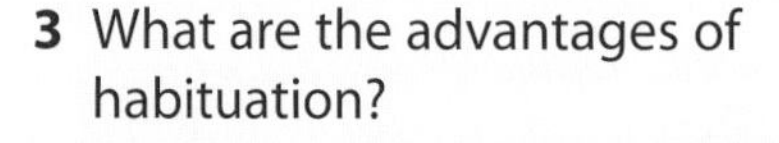

3 What are the advantages of habituation?

4 Write down five examples where you've become habituated to sensory information, for example to a bad smell, feel of something, sounds, sights, tastes.

5 Make a table to summarise the similarities and differences between imprinting and habituation.

Summary Exercise

Higher Questions

Extension Questions

4. How do animals communicate?

By the end of these two pages you should be able to:

- recall that much animal behaviour requires communication
- explain that communication can happen in many different ways – sounds, signals and chemicals (pheromones).

Communication is behaviour that influences the actions of other individuals. Making sounds, giving signals and releasing chemicals called **pheromones** into the environment are some of the ways animals communicate with one another.

Honeybees are social animals, enabling them to work together as a community. Up to 80,000 individuals live together in a colony. The **queen** dominates the colony. She is its only fertile female member. Her job is laying eggs. The other females are the thousands and thousands of sterile **workers**. The hundred or more male bees of the colony are the **drones**. One of them will mate with a new queen.

Bees produce honey from the nectar they collect from flowers. They also pollinate flowers and produce other products such as beeswax, which humans use. Watching honeybees shows us that individual bees:

- usually visit the same type of flower at the same time of day, day after day
- have a sense of direction
- check bees returning to the hive by touching them with their antennae (feelers). If one of the bees is from another hive, they drive it away.

Communication enables the bees of a hive to work together.

A Bees communicate so that only bees from the same hive are allowed to enter.

1 a What is communication?
b What does communication consist of?

Bees searching for flowers find their way to and from the hive using the position of the Sun in the sky. When they return to the hive, the bees tell the other bees where the flowers are by dancing either at the hive entrance or on the honeycombs inside. The type of dance and the angle of the movements tell other bees the distance and the direction they need to fly to find the flowers.

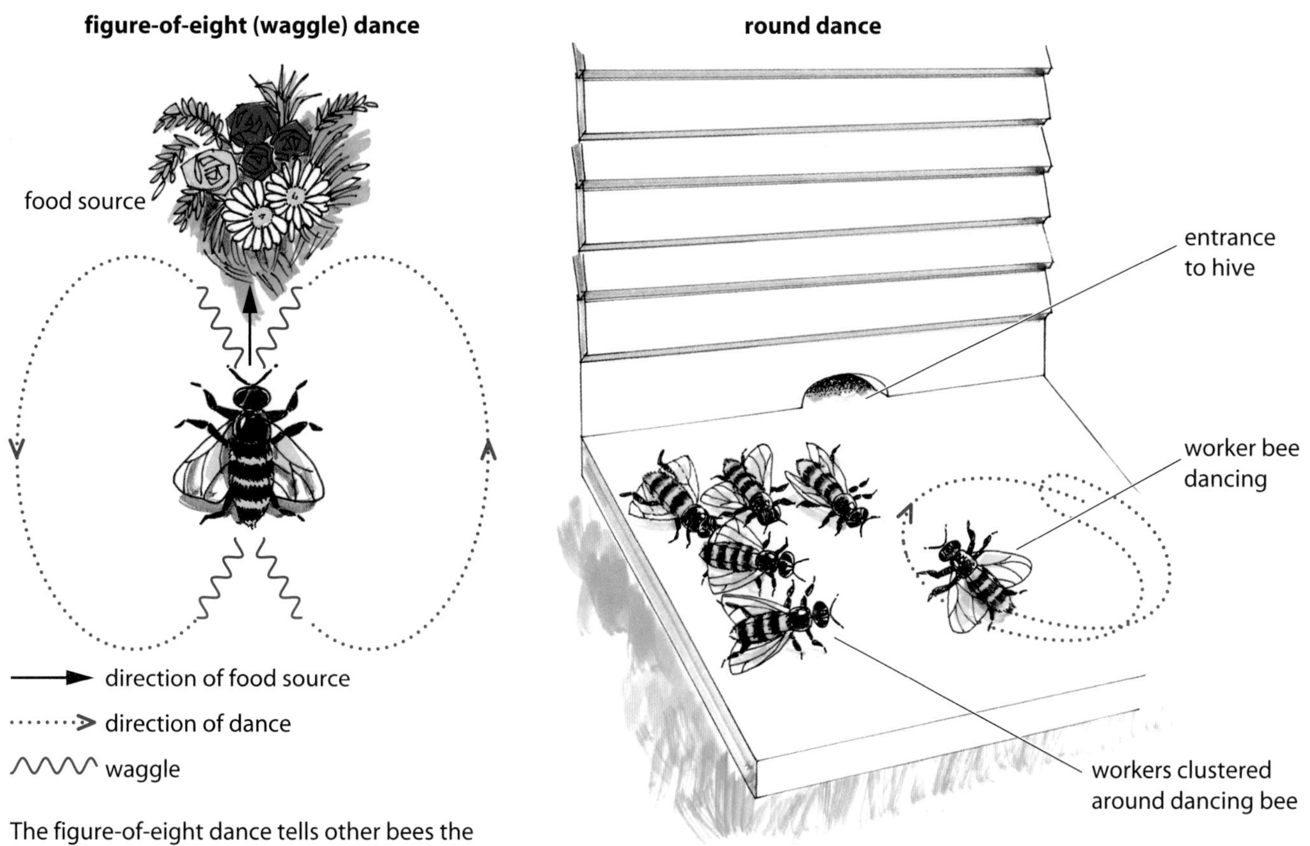

The figure-of-eight dance tells other bees the direction and how far away the food is.

The round dance tells other bees that food is close by.

B Dancing bees: different types of dance communicate the distance flowers are from the hive and their direction.

Honeybees can use pheromones to communicate. For example, the queen produces a pheromone called queen substance which passes to workers when they touch her with their mouth parts. The queen substance:

- tells all the other bees of the hive that the queen is present and in good health
- makes sure that workers remain sterile
- prevents rival queens developing from the eggs the queen lays in the brood area of the hive
- attracts males to her during the mating flight.

Pheromones are also used for communication in many other animal species, particularly in courting behaviour.

2 Explain how different types of dance communicate information about the distance and direction bees must fly to find flowers.

3 What are pheromones?

4 What happens during and after the mating flight of honeybees?

5 Describe different ways in which animals can communicate and give some examples of each.

P

Summary Exercise

Higher Questions

Extension Questions

5. Courting behaviour

By the end of these two pages you should be able to:

- explain that sexual reproduction requires the finding and selection of a suitable mate and can involve courting behaviour.

A How many 'eyes' can you see?

Look at the peacock in photograph A showing off his gorgeous tail. Some of his feathers end in eye-spots.

In the breeding season, wild peacocks establish a territory in clearings in the forest. They attract peahens by calling and then showing off and vibrating their tail feathers when the female arrives. Peahens rarely mate with the first male they see. Usually they will visit a second, third or even fourth peacock before choosing a mate. The size and brilliance of her potential partner's tail and well as the number of eye-spots help the peahen to make her choice.

The way peacocks and peahens interact with one another is an example of **courting** behaviour. To begin with males attract females, then females select a particular mate from a number of possibilities. Attraction through display and **selection** of a mate are all part of courting behaviour. **Sexual reproduction** is the outcome.

In species where males help females rear offspring, their displays are messages about parental duties: providing food, protecting from predators and so on. For example, female barn swallows prefer males with longer tails. A long tail means good health and indicates to the female that the male may be better at carrying out his parental duties.

1 Describe the courting behaviour of peacocks.

2 Using the courting behaviour of peacocks as an example, explain the functions of courting behaviour.

Have you ever wondered?

How does sexual attraction work in humans?

Male woodland birds attract females using **sounds** and songs to advertise their presence. Each species has its own particular sounds so other individuals of the same species will be able to identify the message in the music. To other males an individual's song may mean 'stay away from my territory'; to females his song is an invitation to visit. Threats to rival males and invitations to females are part of the courting behaviour of many different species of songbird.

Some animals produce chemicals which smell. The chemicals are pheromones. For example, a male cheetah may spray urine containing pheromones onto a tree. Other cheetahs passing will know that the area has been 'claimed'. The message in the smell is 'stay away, this is my territory'.

Female silk moths produce pheromones which carry downwind and attract males. Their antennae are so sensitive that males can detect the pheromone several kilometres from the female producing it. The pheromone is a signal to the males that the female is ready to mate.

Once males have picked up the female's scent, they head towards her, following a trail of pheromone which becomes stronger and stronger the nearer they get to the female. Males of other species are not attracted by the pheromone released by the female silk moths. They do not recognise the pheromone's message and ignore it.

Once potential mates have been attracted, one of them needs to be selected. In most cases, females are the selectors. Their choice is guided by the males' displays which attracted them in the first place. The displays probably indicate the health of the males.

3 Describe two functions of pheromones.

4 Describe examples of how female birds judge which male birds are the best mates.

5 Humans show courting behaviour. Describe examples of where humans use display, sounds/song and pheromones to attract a mate.

B The antennae of the male silk moth detect female pheromones.

C A male platy fish (blue) tries to attract a mate.

Summary Exercise

Higher Questions

Extension Questions

6. Mating systems and parental care

P

By the end of these two pages you should be able to:

- explain that some animals mate for life; others select different mates during the mating season
- recall that some animals, in particular birds and mammals, have developed special behaviours, including feeding behaviours, for the rearing of young – they display parental care
- explain that parental care is a successful evolutionary strategy; although it involves risk to the parents, it can increase the chances of survival of the parental genes.

A Part of swans' courting ritual.

Swans mate for life. The behaviour between a male and female helps to cement the bonds which keep them together. **Pair bonding** is reinforced during the breeding season as the parents work together to bring up their offspring.

Red deer do not show pair bonding. In late summer, males move to mating areas where the grass is thick and lush. Plenty of grass means lots of food is available to support the females which gather round them. The males compete with each other for the best places inside a mating area. The males that are best at defending these prime spots mate with the most females.

B Red deer competing.

Animal species where males and females pair for the breeding season or even for life are **monogamous** species. Animal species where an individual mates with a number of opposite sex partners during the breeding season are **polygamous** species.

Monogamy and polygamy are the basic types of sexual behaviour which make up **mating systems**. In monogamous species, both parents are usually involved in looking after offspring. In polygamous species, males do not usually take a role as a parent.

Parental care includes feeding, protecting and helping offspring to interact with others – predators, prey, individuals of the same species, etc. The behaviours help offspring to survive. Parental efforts, however, take time and energy and represent the costs to parents of rearing offspring.

Parental care seems to act against parental interests because of the efforts involved in raising offspring. However, if care increases the chances of offspring surviving, then parents indirectly benefit because their genes pass from one generation to the next. Perhaps parental care is all about increasing the survival of parental genes. It maximises the selfish genetic interests of parents.

Parental care is most common among **vertebrates** such as birds, **mammals** and some **reptiles**. For example, crocodile parents respond to the calls of the young, newly hatched from eggs the female had buried earlier. Both parents dig out the hatchlings, carry them in their mouth to the river and wash them.

C Crocodiles show parental care.

Most parent birds care for their young to a greater or lesser extent. However, cuckoos are an exception. The female lays an egg in the nest of another species. When the cuckoo chick hatches, it pushes the eggs and chicks of the resident species out of the nest. The resident parents think that the cuckoo chick is their own and feed it. The cuckoo parent passes on its genes without putting any time and energy into parental care.

1 What is the difference between monogamy and polygamy?

2 What is 'parental care'?

3 How do pair bonds reinforce parental care?

4 Why does a cuckoo lay its eggs in other birds' nests? Mention parental care in your answer.

5 Describe the differences between a monogamous species such as swans and a polygamous species such as red deer. Mention the following points: time spent together, home-building, competition for mates, parental care.

Summary Exercise

Higher Questions

Extension Questions

7. Feeding behaviour

By the end of these two pages you should be able to:

- explain that feeding behaviours are different depending on the type of food consumed
- recall that herbivores have to eat more food in order to get the nutrients (particularly proteins) they require, so that more time is spent eating
- explain that vertebrate herbivores who feed in large groups usually need to be continually on the move to find new feeding areas
- recall that some animals have developed the use of tools in their search for food.

Hunger is the stimulus which drives the animals in photograph A to find food and eat it. Feeding behaviour is what animals eat, how they find and eat their food, and when they eat it. For example, the antelopes' feeding behaviour is to travel in a large group, eating plant material throughout most of the day. The group is called a **herd**.

A Antelopes feed in a herd.

Animals learn where and when they are most likely to find food. For example, hyenas quickly sense where lions have brought down an antelope. When the lions have taken their fill of the food, the hyenas move in and scavenge the remains. The hyenas' feeding behaviour means they obtain food without the effort of hunting and killing it.

Feeding behaviour varies. For many animals it is a routine. They feed at certain times of the day. For example, female mosquitoes mostly feed at dawn and in the evening. Other animals make the decision to feed depending on their circumstances. For example, rabbits will not feed if they think a fox is nearby.

Cows and sheep are **herbivores** – they eat plants. Herbivores seem to feed all the time. The concentration of nutrients (particularly protein) in the grass is low. So, they have to eat more food to get enough of all the nutrients they need: carbohydrates and oils for energy, proteins for growth and repair, minerals and vitamins.

Have you ever wondered?

Why do cows spend all day eating?

B Wildebeest follow their food.

The wildebeest in photograph B are searching for food. They are following the rains. Where it is raining, grass grows and wildebeest eat. When the rains stop, the wildebeest move onto the next area where it is raining and the grass is growing green and tall. Following the rains to find food is an example of seasonal migration. Moving to new feeding areas means that large groups of herbivores find enough to eat.

1 Why do cows and sheep spend a lot of time eating food?
2 What are the smaller molecules that make up a molecule of protein?
3 Why do wildebeest migrate?

The chimpanzee is using a stick to dig out insects from holes in a log of wood. Is the chimpanzee using the stick as a tool to find food?

Chimpanzees alter the length and shape of sticks, making them better for reaching and probing. Altering the length and shape of its 'food stick' suggests that the chimpanzee is using intelligence and is using the stick as a tool.

C Chimpanzees use tools to help them to find food.

4 What is a tool?
5 How do chimpanzees show that they are probably thinking about their use of tools to search for food?

6 What is the feeding behaviour of:
a a dog
b a mouse
c a flea
d a rabbit?

Summary Exercise

8. Foiling predators

By the end of these two pages you should be able to:

- explain that vertebrate herbivores may feed in large groups or herds, and they may do so for protection in numbers
- explain that herd feeding is a successful evolutionary strategy, even though some members of the herd may be killed
- recall that herbivores have to be good at avoiding, fleeing from, or resisting predation.

A The lions are confused by the stripes, making it difficult to focus on individual prey.

Both tigers and zebras have stripes. The tiger's stripes are camouflage whereas the zebra's stripes dazzle and confuse. Imagine lions creeping up on a herd of zebras. The zebras sense the lions and run in all directions. The lions see a confusion of stripes – which stripes do they charge after? Sick, young and individuals at the edge of the herd are most likely to fall prey to the lions. The rest benefit from safety in numbers and the confusion of stripes.

Have you ever wondered?

Why do fish shoal?

Prey are the animals that predators catch and eat. Living in herds protects the majority of herd members from **predation** although some individuals may be killed. Many pairs of eyes are better than one when looking out for predators. An animal feeding on its own often has to look up to check that predators are not nearby. Looking up means time not feeding. As part of a herd, individuals can spend more time feeding because their companions are also looking out for predators. Being a member of a herd is a successful **evolutionary** strategy. Most of the individuals in the herd survive to produce offspring that inherit the herding instinct. The behaviour, therefore, is reinforced by natural selection and passes down the generations.

1 Briefly describe how most zebras avoid being caught by lions.

2 What is the advantage of being a member of a herd?

Prey animals do not necessarily have to be members of a herd. They can have other characteristics which give them **protection** from their predators.

Camouflage helps prey to avoid capture. For example, stick insects look like plant twigs and are often overlooked by predators.

Shock tactics scare off predators. For example, when threatened, a resting eyed hawkmoth quickly opens its wings to show a pair of huge, brightly coloured eye-spots.

C Shock tactics.

Eyes positioned on the side of the head allow rabbits, antelopes and other prey species to watch for predators approaching from behind as well as to the front.

Long lever-like legs help prey like antelopes to dodge a predator's first rush and run off to safety.

Threatening predators is possible when herd members cooperate with each other. For example, musk ox form a defensive circle to resist an attack by wolves. The more vulnerable animals within the circle are protected by the barrier of horned heads pointing out towards the circling wolves. The larger the herd, the less likely it is that an individual will be the victim of an attack.

B Camouflage.

3 Describe how herbivores can avoid predators by using:
a camouflage
b shock tactics.

4 How can herbivores in herds resist predators?

5 Summarise the different strategies prey species use to foil predators.

P

Summary Exercise

Higher Questions

Extension Questions

9. Catching prey

By the end of these two pages you should be able to:

- recall that some carnivores hunt in packs; others hunt as individuals
- recall that carnivores eat protein-rich food and spend less time than herbivores actually eating
- explain that carnivores have adaptations that enable them to detect and catch their food.

Carnivores eat meat. Meat is a richer source of nutrients, particularly protein, than plant food. For carnivores their problem is catching prey. Once caught, the high quality of the food means that carnivores spend less time feeding than herbivores. Now you know why cats spend most of their time sleeping when not eating!

The adaptations of carnivores to detect, catch, kill and eat prey are illustrated by the following examples.

- The fox's eyes are close together and face the front. The fox has a narrow field of view but can accurately judge the distance to its prey.
- The bear's large, pointed canine teeth grip and tear at its struggling prey. Its cheek teeth cut like scissor blades through flesh and bone.
- The cheetah's thin, long body and flexible spine let its front and hind legs overlap, allowing long strides. A cheetah can achieve bursts of over 110 km/h running in pursuit of its prey.

A A bear's teeth are adapted to tear flesh.

B A cheetah is adapted for speed.

C A tiger is adapted for hunting.

- The tiger's stripes make it difficult to see a tiger among the tall grass where it hunts its prey. The stripes are camouflage which enable the tiger to creep up to its prey undetected.
- Moving to where prey is plentiful increases the chances of predators catching prey. Predators may lurk near a watering hole waiting for the chance to make a kill.

Why do some carnivores hunt in **packs** and others work as individuals? It may be all to do with the size of the prey. For example, a moose is much bigger than a wolf. However, a pack of wolves working together can bring down a moose where a wolf on its own would probably fail.

Feeding in a pack means that there are fewer leftovers from the kill. Scavengers are animals that feed on the remains of prey killed by predators. Packs of carnivores are able to chase off scavengers. This means that each member of the pack has more meat for it to feed on than an animal hunting alone.

1 Why do carnivores not need to eat as frequently as herbivores?

2 Describe how the following characteristics help predators to catch, kill and eat prey:
a eyes
b legs
c teeth
d colouring
e hunting near watering holes.

Have you ever wondered?

Why do dogs greet each other by sniffing?

Most wild cats hunt alone for small prey, leaving few leftovers for scavengers. The hunter eats most of the prey it kills and so pack life would have few advantages.

3 What is the advantage of hunting in packs?

4 What are the advantages of scavenging for food?

5 Use examples to show how hunting in packs and hunting alone are equally successful strategies in different species of carnivores.

Have you ever wondered?

Why are dogs so different from cats?

Summary Exercise

Higher Questions

10. Non-verbal communication

At the end of these two pages you should be able to:

- explain that most mammals are able to communicate their intentions through body posture and facial expression
- recall that facial expressions are species-specific
- explain that a gesture or expression may appear as a threat to one species, but may mean something totally different to another.

Pulling a face is one way of communicating our **emotions** to others. What do the different **facial expressions** of the people in the photograph tell you about how they're feeling?

A Different facial expressions.

Have you ever wondered?

How can people 'read' your face?

Chimpanzees pull faces too. Their facial language is often similar to ours, in appearance and in its meaning. However, emphasis of expression varies. We frown more deeply than chimpanzees, but they have a broader grin. It isn't that humans are sadder than chimpanzees and they are happier than humans. More probably the difference of emphasis is due to differences in the shape of faces. For example, the front of a chimpanzee's face is proportionately broader than a human face.

However, some expressions seem to have different meanings in humans and chimpanzees. In chimpanzees, a full open grin is thought to show fear, rather than happiness.

1 Give five examples of human facial expressions.

2 a Why do humans and chimpanzees have similar facial expressions?
b Why are their facial expressions slightly different?

3 In chimpanzees, a full open grin is thought to indicate intense fear or some other form of excitement. Explain how this shows that facial expressions are species-specific.

B Is this chimpanzee happy?

Can you read body language? Body **posture** and movement tell others a lot about how you are feeling and your intentions. For example, if you sit towards someone, you are probably interested in the person and want to know more. People often use their hands to communicate information – the good-bye wave, the victory clenched fist, the circle made with thumb and finger which usually means 'OK'. These are called **gestures**. A gesture may not mean the same thing in all cultures. For example, in France the 'OK' gesture means something is worthless, and in Greece it means something very rude!

Mammals, birds, reptiles and many other types of animal use facial expressions and body language to convey information to other individuals. Does a cat rubbing its sides against your legs means it loves you? As far as the cat is concerned it's marking you as part of its territory. Scent glands on the sides of its face and the base of the tail mark you as its own.

When a cat runs to greet a friendly animal, its tail will be held high. The raised tail indicates happiness and a welcome. A cat swishing its tail spells trouble. Generally the cat is cross or feeling threatened. Add an arched back with bristling fur to a tail lashing to and fro and the cat is really angry.

C The cats disagree – they use body language, loud yowls and scent. The confrontation will end either in a fight or with the less dominant cat running away.

A dog wagging its tail gives out an entirely different message. The dog is happy. However, a tail down or tucked under the body are signs of fear. The language of the tail is different for cats and dogs. However, fur standing on end means much the same thing in either case. The raised hair makes the animals' body look bigger and tells any other animal to beware.

4 Briefly describe the effect of culture on the meaning of hand gestures.

5 Does a cat rubbing against your legs mean it loves you? Explain your answer.

6 Describe how the message of body posture and tail language is different in cats and dogs.

P

Summary Exercise

Higher Questions

Extension Questions

11. Being human

At the end of these two pages you should be able to:

- explain that humans have developed highly complex ways of communicating
- recall that humans transmit knowledge of past events, emotions and complex ideas to other humans.

Like other mammals, humans can communicate through posture, sounds, scents, gestures and facial expressions. However, humans also have language, writing and numbers, which means that we can communicate with people in other places, and with future generations.

Language is spoken and written. Spoken language developed long before written language, but scientists are not certain how speech began. Humans have features not found in other animals which make speech possible. These features include a voice box which amplifies sounds produced by air passing through the vocal cords and a brain which can interpret the sounds. The sounds are symbols we use to describe current events (what is happening now), events which happened in the past (history) and events which might happen in the future.

Speech is generally recognised as the biggest difference between humans and other animals. Chimpanzees, dolphins and whales make sounds which are probably different forms of spoken language. However, human speech is more complex and allows **self-awareness**, highly developed emotions and personal memories.

Scientists cannot agree when humans first began to use spoken language. Cave paintings have been found from 40 000 years ago. Perhaps speech arrived around the same time. However, the fossil record might suggest an earlier date. The pattern of marks on the sides of fossil skulls of an early type of human called *Homo erectus* indicates that the area of the brain which enables humans to speak developed 1–1.5 million years ago.

Human speech does not leave a record unless it is written down. Ancient symbols which represent thoughts and ideas understood by other people have been dated to 7000 years BCE. The symbols are not writing as we understand it but are a form of proto-writing. The discovery of the Tărtăria tablets from the 5th millennium BCE suggests how earlier symbols might have developed into systems of writing. The Tărtărian symbols are in rows and lines, giving the impression of an organised text. Unfortunately little is known about their meaning, making it difficult to pin-point when writing emerged from proto-writing.

1 What human features make speech possible?

2 Which abilities might possibly make humans different from other animals?

A These stones are one of the oldest known examples of written communication.

Writing systems are broadly based on pictures or an alphabet of letters. The pictures are arranged in a variety of ways rather than in horizontal lines of letters. Each picture and its arrangement represents a word or part of a word; each letter a sound or combination of sounds. The words and sounds have meanings which are understood by other people.

Who thought of numbers as a form of written language? The numerals 0–9, combined as numbers, can be used for counting (e.g. there are 14 plates on the table) and ordering (e.g. this is the 21st century). Probably the ancient Egyptians and the people of Babylon (in modern-day Iraq and Iran) came up with the idea. A stone carving from Egypt dating from around 1500 BCE shows a combination of numerals for the number 4622.

B Communication using numbers.

3 Briefly describe the two forms of writing system.

4 How may numbers be used to communicate?

5 Summarise the ways humans transmit knowledge of past events, emotions and complex ideas to other humans.

Summary Exercise

Higher Questions

12. Self-awareness

At the end of these two pages you should be able to:

- explain that humans are conscious of the outcomes of their actions, and as a result are more self-aware than other animals.

It was the Greek philosopher Socrates (469–399 BCE) who said 'know thyself'. His statement suggests that we know we exist. We are self-aware. The notion of self-awareness also includes the idea that each of us exists as an individual, with our own thoughts, separate from other people. It also includes the idea that we are **conscious** of the outcomes of our actions.

We humans can recognise ourselves in a mirror. Looking in a mirror for a mark on your back suggests that you are aware of your body image. The mirror test is a test for self-awareness that is used to discover if an animal can recognise its own reflection as an image of itself. Chimpanzees, orangutans and dolphins may recognise themselves in mirrors.

A The mirror test for self awareness – you can recognise yourself.

Like humans, dolphins have a big brain. Their ability to understand language (a complex repertoire of whistles, trills, grunts and other sounds) and remember information is as good as that of monkeys and apes.

1 What is self-awareness?

2 Briefly explain how the mirror test suggests that an individual is self-aware.

Dolphins love to bow-ride – swimming just to one side of the bows of a ship so that they can surf the ship's bow wave. The dolphins are playing for no better reason than they love doing it. The behaviour is different from the play behaviour of lion cubs. Lion cubs' play is helping then to learn how to hunt. Dolphins bow riding seem to be consciously playing just for play's sake.

B The Thinker. Humans are self-aware.

Self-awareness allows us to have an idea of what other people may be thinking. We might recognise another person's mental state and use our understanding to influence their behaviour to our advantage. Watching the behaviour of chimpanzees underlines the idea. For example, a young chimpanzee sees some food not noticed by older and more dominant individuals. The youngster gives an alarm call. The other chimpanzees are distracted because they think a predator is lurking nearby. Meanwhile the youngster swiftly makes off with the food and eats it.

The young chimpanzee's behaviour suggests that it was aware of using a false alarm call to affect the behaviour of the other chimpanzees. These ideas support the idea that the behaviour of some animals shows self-awareness – that they are aware of the outcomes of their actions. However, not everyone agrees.

Self-consciousness is to be self-aware but with a difference. To be self-conscious means that you feel *too* aware of even the smallest of your own actions. It can affect your ability to carry out complex actions. You miss your shot at an open goal or you stumble over the fingering of your favourite guitar riff which you played perfectly in the privacy of a rehearsal room. In each case the watching crowd puts you off because you are self-conscious about being the centre of attention.

3 Are humans the only self-aware animals? Give reasons for your answer.

4 What does it mean to be self-conscious?

5 How do you know that humans are self-aware?

P

Higher Questions

Extension Questions

13. Family society

At the end of these two pages you should be able to:

- recall that humans have developed from small family groups of hunter-gatherers, to complex societies capable of (gross) modification of their own environment.

About 12 million people lived in the world 10 000 years ago. Today, the world population is 6.5 billion and rising.

Everyone needs food and a home. Many people want a wide range of manufactured goods like televisions and cars. Increasing population size increases the demand for the raw materials needed to make these goods. This can affect the environments which are the source of the raw materials.

Humans' impact on the environment increased when people shifted from a nomadic, hunting and gathering way of life to a more settled lifestyle based on permanent communities. Before then, the human impact on the environment was no more than that of other medium-sized animals. Now we live in cities, transform the landscape and have the nuclear technology to wipe out life on earth.

1 How does an increasing population affect the environment and why?

2 Briefly describe how humans have an impact on the environment.

Farming takes up 19 million hectares of land in the UK.

Clearing tropical rainforest for cattle ranching quickly exhausts the nutrients in the soil. Semi-desert develops.

human impact on the environment

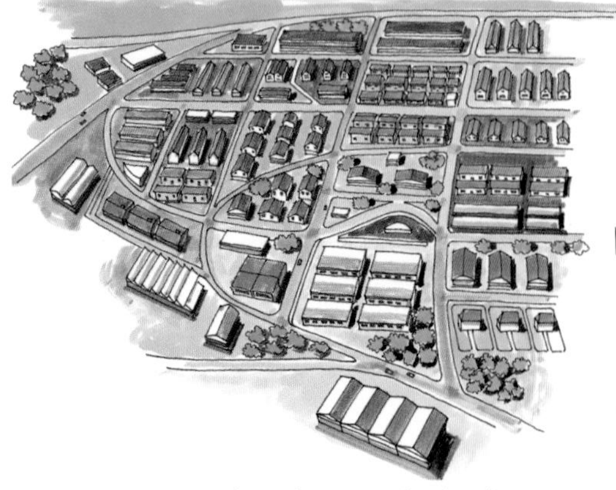

Housing and roads cover the landscape, destroying environments and wildlife.

Mining extracts coal and minerals from the land. Craters and heaps of waste material disfigure the landscape.

A Human impact on the environment.

About 60 000 years ago, humans lived as hunter-gatherers. Family groups set up camp for a few days while animals were hunted and wild plant food was gathered in the local area. Families moved from campsite to campsite, perhaps following the animals and the seasonal rains. The rains encouraged plant growth and so meant an abundant supply of food.

About 10 000 years ago, some hunter-gatherer communities settled in an area, harvesting wild wheat and other wild grasses. Families could store seed grains meaning they did not have to move from place to place in search of food. They had a more settled way of life, enabling them to live in permanent communities.

By 2500 years ago, farming was an established way of life. Animals and plants with characteristics which were especially useful to us were selected and bred. Food was more plentiful and not everyone was needed for farm work. Some people were able to specialise in weaving and pottery, others trained as priests, doctors and lawyers. Large towns and society emerged.

Today, people mostly live in big cities. Food is provided by relatively few people, working in farming. Industry and technology offer improved living standards and increased leisure time. Mass transport allows people to travel around the world. Societies which developed over centuries in different parts of the world are now in touch daily through travel and the internet.

B Hunter–gatherers finding food in the local area.

C City life.

3 Summarise the lifestyle of hunter-gatherers.

4 Why did farming enable people to live in permanent communities?

5 Briefly explain how society might have developed from small family groups of hunter-gatherers.

P

Summary Exercise

Higher Questions

Extension Questions

14. Humans and other apes

At the end of these two pages you should be able to:

- recall that humans are one of the great apes, who have all developed from a common ancestor.

Humans are one of the **great apes**. Let me introduce you to our cousins. Our 'cousins' in the sense that humans, chimpanzees, gorillas and orangutans shared a common ancestor 6–6.5 million years ago.

A The great apes.

Evidence from comparing the genetic material of the great apes shows that humans, chimpanzees (including **bonobos** – pygmy chimpanzees) and gorillas are closer than any of us are to orangutans. Also, humans and chimpanzees are the closest cousins of all. About 99% of our genetic material is the same.

In the summer of 1960, Jane Goodall left England for Tanzania to study chimpanzees in the wild. At the time very little was known about their behaviour, social structure and daily life, and to begin with the chimpanzees ran away when Jane approached. After a few months the chimpanzees accepted her presence. The breakthrough came when the chimpanzee Jane called David Greybeard risked visiting Jane's camp. Spotting bananas on the camp table, he snatched them up and hurried away with the prize into the forest. Soon, other chimpanzees entered the camp and Jane could begin to watch them at close quarters.

B Jane Goodall with company.

Jane was the first to watch chimpanzees use twigs as tools and cooperate to hunt monkeys. Over the years Jane's work has shown that chimpanzees are like humans. They cooperate with each other, make tools, solve problems, quarrel between themselves and comfort one another. Their emotions and ours are similar. Jane's work on understanding chimpanzees helps us to understand human behaviour.

An American, Dian Fossey, was murdered in Rwanda in 1985. Her death remains an unsolved mystery. Over thousands of hours Dian Fossey earned the trust of gorillas, allowing her to watch their behaviour closely. She was curious about their lives and determined to protect them from poachers, who killed the gorillas to supply souvenirs for the tourist trade and the growing market for **bush meat**. Her work has helped to conserve gorillas, but her aggressive behaviour towards poachers was probably the reason for her death.

C Dian Fossey – the first friendly gorilla/human contact recorded.

Hunting for bush meat continues to threaten the survival of the great apes in the wild. Other threats include civil war, and clearing forests and bush country where the great apes live to make way for farming. The work of Jane Goodall, Dian Fossey and the organisations they founded helps to publicise the problems facing the great apes. Reducing demand for bush meat, protecting the great apes' habitat and highlighting the place of humans as just one of the four great apes contribute to the ongoing efforts for their conservation.

1 What evidence suggests that chimpanzees are humans' closest cousins among the great apes?

2 List some of the characteristics of behaviour shared by humans and chimpanzees.

P

3 Briefly explain Dian Fossey's work.

4 List some of the human activities which threaten the survival of the great apes.

5 Why is it difficult to save the great apes from extinction?

Summary Exercise

Higher Questions

Extension Questions

15. Using animals

At the end of these two pages you should be able to:

- explain that humans have domesticated animals to help them hunt and have exploited herd animals to provide a constant and dependable source of food
- recall that humans have exploited animals in other ways, as a source of clothing and domestic materials and, more recently, for medical purposes
- recall that humans also use animals as a source of entertainment and companionship.

Animals entertain us, they are our companions as pets, but we also eat them and use them in many different ways.

Some animals get used to being with people more easily than others. Human control of their breeding and living conditions eventually led to their **domestication**. They became tame.

We domesticated animals for what they could do for us. Dogs, bred from wolves, helped us to guard our homes and to hunt. Cats kept the rats and mice that raided our food stores under control. Horses were attached to carts about 5000 years ago, and ridden when hunting and into war. Herd animals were used as a source of food.

A Knights used horses for medieval warfare.

Sheep, pigs, goats and cattle were first domesticated about 10 000 years ago. Bringing them into human care provided a dependable source of meat, milk and fibres. Domesticating animals (and harvesting grain) enabled humans to shift from a hunter-gatherer lifestyle to a more settled way of life based on permanent communities. Now animals are also used as pets. Pets offer companionship.

We do not have to work as hard in our relationship with a pet as we do with other humans. Our reward is love, gratitude, fun and company. Pets help to prevent feelings of loneliness and isolation, and children caring for their pets learn to care for something other than themselves.

Some pets combine companionship with work. For example, guide dogs are eyes for the unsighted; 'sniffer' dogs help the police to track people suspected of crimes and to find drugs and explosives. Horse riding improves the skills of the disabled.

Animals can be entertaining. A dolphin's tricks amuse us, horse racing thrills us, the animals used in films and television are part of the story unfolding before our eyes, and the animals held captive in zoos and circuses delight us with their antics. Also, we might feel that zoos are doing a good job conserving animals whose habitat has been destroyed.

B Zoos can help to conserve endangered species.

Most of us benefit from modern medicines. Experiments on animals are often used as part of scientific research into the production of modern medicines. Most of the experiments are carried out on rats and mice.

Many people believe that using animals to test new medicines is necessary to help doctors know how best to treat diseases. However, alternative methods should be used wherever possible.

1 Briefly explain why and how humans first domesticated dogs and cats.

2 Why did the domestication of animals and harvesting plants allow humans to develop a settled way of life?

3 Give three reasons why people keep pets.

4 Describe how animals can be used for entertainment.

5 Make a table showing how humans use other animals. In each case say whether you agree with the use, and give a reason.

P

P

Summary Exercise

Higher Questions

Extension Questions

16. Issues about animals

At the end of these two pages you should be able to:

- explain that it is a mistake to interpret behaviour observed in other animals as showing human characteristics (anthropomorphism)
- explain that it is also a mistake to assume that human and animal behaviours have nothing in common
- explain that humans now debate the ethics of the use of animals; some people consider that animals have rights comparable or identical to human rights.

Many people like watching cartoons and playing computer games, especially where animals are the main characters. It's all make believe, of course. However, you are caught up in the stories because you think the behaviour of the animal characters shows human characteristics. **Anthropomorphism** describes the way humans pin their thoughts and behaviour onto animals and objects. For example, your cat brings a mouse into the house, so you tell it off and it walks away. You might say the cat is 'sulking', a human characteristic, when it is probably moving off because you shouted at it.

The difference between humans and other animals is only a matter of degree. We all share common ancestors. Humans share a common ancestor with the other great apes. So it is not surprising that humans share similar patterns of behaviour with chimpanzees.

A Is this cat wickedly sneaking up on the mouse? Or is catching mice just part of a cat's nature?

1 When studying animal behaviour, why should you not describe the behaviour in terms of human characteristics, such as jealousy, bossiness and modesty?

2 Charles Darwin (1809–1882) argued that differences in behaviour between humans and other animals is only a matter of degree. Do you agree? Give reasons for your answer.

Have you ever wondered?

Do animals have rights?

In the past, we used animals without considering their well-being. Now we have laws to protect pets from cruelty and to prevent suffering in animals being prepared for slaughter. People are more aware that animals have rights – but do animals have the same rights as humans to health, safety and well-being?

Battery hens are reared intensively. They live in small cages where they cannot stretch their wings. They may peck themselves in frustration. The behaviour is prevented by removing the tips of the birds' beaks, called de-beaking. Many people think that rearing animals intensively is cruel and they try to buy free-range eggs and meat.

Animals penned up in cages in zoos and circuses show behaviour suggesting boredom and frustration. The big cat pacing up and down in a cage on its own is almost certainly an animal in distress.

Is it right to use animals to test medicines? Most people benefit from modern medicines. However, some people think that benefits are not the issue. For them it is a question of **morals** (what is right or wrong) and ethics (the action we take as a result of our moral judgement). They think that using animals to test the effectiveness and safety of new medicines is cruel and wrong.

B Some people defend animal testing.

Finding alternatives to animal testing is not easy and takes a long time. There are many problems in discovering how effective or safe a new medicine might be. Sometimes animal testing is the only available method and scientists have to find answers to difficult problems.

3 Briefly explain why people might think that rearing animals intensively for their meat and other useful products is cruel.

4 Should we test the effectiveness and safety of new medicines on animals before the medicines become available to us?

5 Do you think that anthropomorphism affects our views on animal rights? Explain your answer.

Summary Exercise

Higher Questions

Extension Questions

P

17. Questions

D

P

1 The offspring of mammals and birds are often cared for by their parents. Parental care includes feeding and protecting offspring. The care often seems to be to the parents' disadvantage. Explain how parental care has become a successful evolutionary strategy. Use the words selection, selfish, evolution and gene in your explanation.

2 Peacocks show off their gorgeous tail feathers to peahens. Which peacocks are chosen as mates by peahens depends on different features of the tail feathers. Suggest two features which might influence a peahen's selection of a mate and explain why they might influence her decision.

3 Why do songbirds sing?

4 Woodlice move slowly or stop moving altogether in damp conditions. Explain how their behaviour increases their chances of survival.

5 Different environmental factors affect the behaviour of animals. Copy the two columns below. Match each environmental factor to its effect on behaviour by drawing lines between the two columns. One has been done for you.

Environmental factors	**Behaviour of animals**
light/dark	single-celled plant-like organisms move towards weak acids
wet/dry	birds nesting in spring
heat/cold	fish avoid water which has a low oxygen concentration
chemicals (pH)	owls fly at night
oxygen concentration	woodlice seek damp places
length of day	hedgehogs hibernate under a pile of logs

6 Explain the statement, 'In classical conditioning the response to a primary stimulus is unconditioned but can be conditioned by a secondary stimulus.' Use the word 'associate' in your explanation as well as some of the words in the statement.

7 What information does a worker honeybee communicate to other workers when it performs a:
a round dance
b waggle dance?

8 Using different examples, explain how chemicals called pheromones, released by individuals into the environment, affect the behaviour of other individuals.

9 Zebra feed on the grass of the savannah (grassland) of East Africa. The seasons (wet and dry) affect the growth of grass and therefore the amount of food available to zebra. Large herds of zebra migrate across the savannah looking for grass. Lions in groups hunt zebra.
a Suggest *two* advantages to zebras of living in herds.
b Why do herds of zebra need to migrate?
c Explain why it is more efficient for lions to hunt zebra as a group rather than on their own.
d A zebra's eyes are positioned on either side of its head; a lion's eyes face forwards:
i suggest an advantage to the zebra of having its eyes positioned on either side of its head
ii suggest an advantage to the lion of its eyes facing forwards.

10 Humans and chimpanzees have facial expressions in common. Grinning is an example.
a Explain why humans and chimpanzees have facial expressions in common.
b Why might it be a mistake to assume that a chimpanzee is happy when it is grinning?

11 The Russian scientist Ivan Pavlov was the first to study conditioned responses scientifically. He noticed that when food was placed in a dog's mouth, the flow of saliva increased. He also noticed that the flow of saliva increased as soon as the animal smelt his hand, even before the food was placed in its mouth. The salivary reflex was made stronger by following Pavlov's personal smell with the taste of food. After a period of presenting the dog with both the personal smell and the taste of food, the personal smell alone was enough to make the dog produce as much saliva as if it were given food.

Use the information in the passage, and your own knowledge, to answer the questions.

a Why did Pavlov use the word 'conditioned' to describe the dog's responses?

b Which is the primary stimulus for the dog's response?

c Which is the secondary stimulus for the dog's response?

d What is meant by the phrase 'the salivary reflex'?

e Trial-and-error learning is another type of conditioning. How is it different from the conditioning described in the passage above?

12 Why might it be a mistake to interpret behaviour observed in animals as showing human characteristics?

13 Explain why some people consider that animals have rights comparable with human rights.

14 Explain how humans have developed historically from small family groups of hunter gatherers to modern complex societies.

18. Glossary

D

***anthropomorphism** The way we describe animals, plants and objects as having human characteristics.
***behaviour** The responses of animals to what is going on around them.
***bird** A warm-blooded vertebrate covered with feathers whose fore-limbs are wings.
bonobo One of the two species comprising the chimpanzee genus (usually called the pygmy chimpanzee or dwarf chimpanzee).
bush meat The meat of wild animals which live in tropical woodlands, especially in African countries.
***carnivore** An animal which only eats meat.
***communicate** To pass information between individuals.
***conditioning** The process where animals learn to associate a desired outcome with a stimulus that is not directly linked to the possibility of the outcome.
***conscious** Aware of one's own existence, sensations, thoughts and surroundings.
***courting** Behaviour which includes the attraction and selection of a mate.
domestication The process of bringing animals under human control. The animals become tame.
drone Male honeybee.
***emotion** A feeling, such as love or fear, that causes an aroused state of mind, increased heart rate and breathing rate.
***evolution** The theory that present-day living things are descended from ancestors that have changed through thousands of generations.
***experience** The total of the things that have happened to an individual and of his/her past thoughts and feelings.
***facial expression** Movement of facial features (e.g. eyes, mouth, nose, forehead) which indicate a person's mood and feelings.
***gesture** The use of hands to communicate information such as the expression of feelings.
***great ape** A member of the group which includes humans, chimpanzees, gorillas and orangutans.
***habituation** The process where an animal no longer reacts to a stimulus which caused a response to begin with.
***herbivore** An animal which only eats plants.
herd A group of mammalian herbivores.
imprint The process where young animals bond with animals (including humans) who are not necessarily their parents. Young animals may also imprint on moving objects.
***inherit** In biology, the process where characteristics controlled by genes pass from parent(s) to offspring as a result of reproduction.
***instinctive** Split-second behaviour which is automatic and not thought about.
***learning** Behaviour which changes in the light of experience.
***mammal** A warm-blooded vertebrate covered with hair. Females feed their offspring on milk produced by mammary glands.
mating system How individuals behave towards one another for breeding and how different behaviours affect the way individuals are organised socially.
monogamous Where males and females pair for the breeding season and even for life.
morals The values and reasons (good or bad) which inform us about the decisions we make.
natural selection The process that favours the individuals in a group with the characteristics that best suit them to survive.
***pack** A group of mammalian carnivores.
pair bonding The behaviour which forms bonds between a male and female animal of the same species, keeping them together.
***pheromones** Chemicals released by an individual into the environment, affecting the behaviour of other individuals.
polygamous Where one individual mates with a number of opposite-sex partners during the breeding season.
***posture** Positions of the body which can communicate an individual's attitude to other individuals.
***predation** The act of an animal (the predator) killing another animal (the prey) for food.
***protection** Defence against hostile individuals.
queen The only fertile female in the honeybee nest. All the members of the nest are her offspring.
reptile A cold-blooded vertebrate covered with horny scales which restrict the loss of water from the body.
***selection** Where one individual is preferred to another.
self-awareness The idea that we know that we exist.
***sexual reproduction** Process involving two parents, male and female, to produce a new individual. Sex cells fuse, called fertilisation, and develop into a new individual.
***signal** A sign which triggers appropriate behaviour in another individual.
***sound** Form of energy transmitted by the vibration of air.
***vertebrate** Animal with a backbone (vertebral column).
worker Sterile female offspring of a queen honeybee.

*glossary words from the specification

C3.3

Chemical detection

A Modern chemicals.

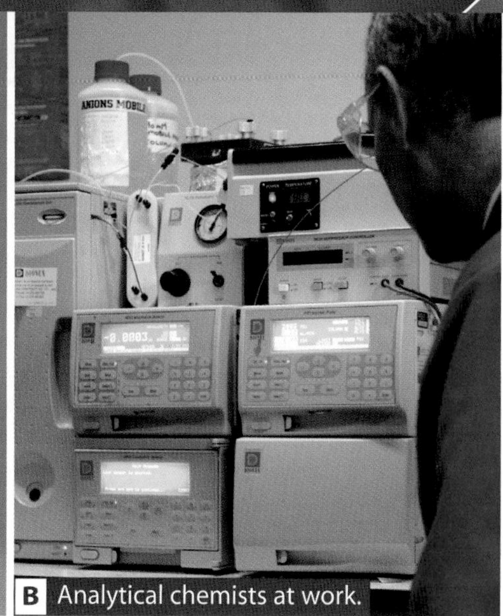
B Analytical chemists at work.

Can you imagine a world without modern medicines, plastics, soaps, detergents, textiles, cosmetics, paints, food additives, fertilisers, solvents, pesticides and all the other products of the chemical industry? These chemicals help to improve everybody's standard of living, and analytical chemists play a central role in their manufacture.

As a chemist you are often asked to analyse things. What's this stain on my shirt? What's in this liquid? What's this white powder, I found it in the garage? These questions are always easier to ask than they are to answer. Even the most ordinary of substances can be complex mixtures which need a team of trained analytical chemists and a well-equipped laboratory for a full analysis.

Analytical chemists work in many places, including forensic science, consumer advice, environmental protection, health care, quality control and scientific research. This topic introduces you to some of the skills and techniques used by all kinds of analytical chemists.

In this topic you will learn that:

- cations and anions are present in many samples and can be identified
- amounts of substances present can be calculated in moles
- you can calculate the amount of raw material needed in a chemical reaction to produce a certain mass of product
- it is important to know the purity of a substance and that different users require different levels of purity.

Look at these statements and sort them into the following categories:

I agree, I disagree, I want to find out more.

- Our tap water is pure.
- All metal ions produce coloured flames.
- Acids and alkalis are easy to identify.
- All ionic substances are soluble in water.
- The same volumes of all gases have the same mass.
- It is easy to find the amount of acid in a solution.
- Chemists always measure amounts in grams.
- There is a simple single test for every substance.

1. Water, water everywhere

By the end of these two pages you should be able to:

- explain the importance of a reliable, pure water supply in everyday life
- describe how water is used in everyday life and explain why it is important not to waste it
- explain why substances need to be identified and their purity determined
- explain the difference between qualitative and quantitative analysis.

We all need a safe supply of water to survive. Over 70% of the human body is made up of water, and health experts recommend that we drink at least 2 litres of water a day. Although water seems to be everywhere – in our oceans, seas, lakes, and rivers – most of it is unfit for drinking. Each year millions of people die because they don't have a safe supply of drinking water. It is therefore very important that we look after our sources of water and that we don't waste water, as the consequences can be severe.

We need water for cooking, cleaning, washing, sanitation, farms and industries and of course for drinking. We are fortunate, in this country, that clean safe water is provided to almost every home. Water is treated by screening and filtering to remove solid particles, chemical treatments to eliminate impurities and chlorination to kill bacteria.

A Water is filtered to remove larger particles of solid matter.

1 Why is most of the water in seas and lakes unfit to drink?

2 Describe three uses of water in the home.

Have you ever wondered?

How pure is our water and how pure does it need to be?

Analytical chemists check the **purity** of the water throughout its treatment. There are two main types of analysis. **Qualitative** analysis, which investigates the kind of chemical present in a sample and **quantitative** analysis, which measures the amount of each chemical present.

B Analytical chemists check the purity of our drinking water.

3 What do analytical chemists do in a company that monitors food safety?

4 What are the two main types of analysis?

The purity of water is checked against government standards, which set maximum limits for chemicals and microorganisms. Our standards are high and ensure that water is not only safe to drink but looks and tastes good as well.

5 Why are government standards for water purity needed?

C Analytical chemists check the water from natural sources and factory waste.

Water comes from natural sources, so water pollution is a major concern. Pollution can come from a number of sources. Industry, which uses large amounts of water, can discharge dangerous chemicals in its waste. Fertilisers used by farmers can seep into rivers and lakes. Indeed we can all be responsible at times, using detergents and other household chemicals which are then washed down the drains.

Many countries have now passed laws to control the discharge of industrial and domestic waste. In the UK, water sources are tested for pollution by analytical chemists. This careful monitoring of our environment has greatly improved the cleanliness of our water.

6 **a** What do farmers use that can cause water pollution?
b Name three household chemicals which could cause pollution problems.

7 Explain how analytical chemists help produce a reliable water supply, and why it is so important.

Higher Questions

Extension Questions

2. Testing for ions

By the end of these two pages you should be able to:

- explain how ionic substances are identified by using tests for each type of ion they contain
- explain why the test for each ion must be unique
- describe and explain the tests for: sodium (Na^+), potassium (K^+), calcium (Ca^{2+}) and copper (Cu^{2+}) ions, using flame tests.

Tap water and bottled water always contain dissolved compounds. Some of the compounds in the water contain **ions**, atoms or groups of atoms, which have lost or gained electrons. These compounds are called **ionic substances** and they are usually formed between metals and non-metals. Ions are tiny charged particles, with either a positive or a negative charge. Metals usually form positive ions, called **cations**, by losing electrons. For example sodium atoms form positive ions when their atoms lose one electron.

$$Na \rightarrow Na^+ + e^-$$

Non-metals usually form negative ions, called **anions**, by gaining electrons. For example chlorine atoms form negative ions by gaining one electron.

$$Cl_2 + 2e^- \rightarrow 2Cl^-$$

To find out which ionic compounds are dissolved in drinking water, we need to use at least two tests, one for each of the ions present. Photograph B shows some typical ionic compounds. Although some are coloured, most are colourless crystalline solids, and it is impossible to identify the ions present by appearance alone. To analyse ionic substances, a series of tests is needed. Each of these tests must give a positive result with only one ion. The tests must be unique for each ion so you can be sure of the conclusions of the analysis.

B Some common ionic compounds and their ions.

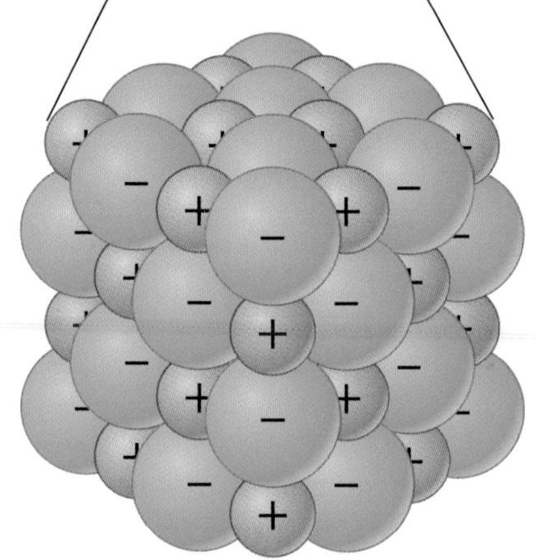

A Sodium chloride is a common example of an ionic compound.

1 What types of element usually form an ionic compound?

2 a What are cations?
b What are anions?

3 Why is it important to have a unique test for each ion?

4 Why can you not use appearance as a test for ions?

Fireworks contain a mixture of chemicals. Some send them up into the air, some make them explode and some give them colour. The colour usually comes from small amounts of certain metal ions which produce different coloured flames.

C The colours of fireworks are due to different metal ions.

Element	Flame colour
calcium (Ca^{2+})	brick red
sodium (Na^{+})	yellow
potassium (K^{+})	lilac
copper (Cu^{2+})	green

D Metal ions produce different coloured flames.

5 What element produces the flame colour in photograph E?

Flame tests were one of the simplest and first qualitative analysis techniques to be used. If a clean piece of wire is dipped in a **solution** of a compound and then held in the hot part of a Bunsen flame, the colour produced can indicate the metal ions present.

The colours are produced as electrons in the atoms move between the energy levels. Different metals have different electron arrangements, with different energy levels, therefore different metals can give out different colours of light.

Other examples include lithium and strontium ions, which produce a bold crimson coloured flame. However, the colours of these two are far too similar to be used as a test for the ions.

E Flame testing.

6 **a** Explain why flame tests are a qualitative method of analysis.
b Why are flame tests less useful for identifying lithium ions?

7 Why must at least two tests be carried out to identify an ionic compound?

P

P

Summary Exercise

Higher Questions

Extension Questions

3. Looking for cations

By the end of these two pages you should be able to:

- describe and explain the tests for: aluminium (Al^{3+}), calcium (Ca^{2+}), copper (Cu^{2+}), iron(II) (Fe^{2+}), iron(III) (Fe^{3+}) and ammonium (NH_4^+) ions using sodium hydroxide solution
- explain how precipitation reactions can form the basis of some tests for ions.

Water from rivers and lakes can contain a number of ions. In hard water areas, where there are chalk or limestone rocks, the water contains calcium ions (Ca^{2+}). In some areas copper ions (Cu^{2+}) can be present due to industrial pollution.

A Water from rivers and lakes will contain many dissolved substances.

Flame tests are a good starting point for any analysis but other tests are needed. In fact, a series of tests is often required to find a unique set of results for each ion.

The next test usually involves adding a few drops of sodium hydroxide to a solution of the unknown sample. Many metal hydroxides are insoluble so a **precipitation** reaction can occur and the **precipitate** formed can identify the ion present.

Photograph B shows the results of adding sodium hydroxide to a number of different ion solutions. The results are summarised in table C.

B Precipitation can be used to identify some ions.

A solution can often be obtained by simply dissolving the solid in pure water. However, if it is insoluble in water, then hydrochloric acid and nitric acid can be tried as possible solvents.

1 Why can we not use flame tests to detect all cations?

2 If an ionic solid won't dissolve in water, what other solvents could you try?

Cation	Symbol	Precipitate
ammonium	$NH_4^+(aq)$	none
aluminium	$Al^{3+}(aq)$	white
calcium	$Ca^{2+}(aq)$	white
copper(II)	$Cu^{2+}(aq)$	blue
iron(II)	$Fe^{2+}(aq)$	green
iron(III)	$Fe^{3+}(aq)$	brown (rust)

C Effect of adding sodium hydroxide solution.

The precipitate will be a solid hydroxide of the cation. For example, if the compound was copper(II) sulphate, the precipitate would be copper(II) hydroxide. If the unknown sample was iron (II) nitrate, then the precipitate would be iron (II) hydroxide. The **balanced equations** for these reactions are given below.

$$CuSO_4(aq) + 2NaOH(aq) \rightarrow Cu(OH)_2(s) + Na_2SO_4(aq)$$

copper(II) sulphate + sodium hydroxide → copper(II) hydroxide + sodium sulphate

BLUE

$$Fe(NO_3)_2(aq) + 2NaOH(aq) \rightarrow Fe(OH)_2(s) + 2NaNO_3(aq)$$

iron(II) nitrate + sodium hydroxide → iron(II) hydroxide + sodium nitrate

GREEN

Table C shows that further tests are needed for ammonium, aluminium and calcium ions. This is because ammonium ions do not form a precipitate, and both aluminium (Al^{3+}) and calcium (Ca^{2+}) ions form a white precipitate. However, aluminium and calcium can be distinguished by adding excess sodium hydroxide. In excess sodium hydroxide the calcium precipitate is not changed, but the aluminium precipitate starts to dissolve.

Like the alkali metal ions, ammonium ions (NH_4^+) form no precipitate with sodium hydroxide solution. To test for the ammonium ion, heat the unknown sample with concentrated sodium hydroxide. If the ions are present the smelly alkaline gas called ammonia will be given off. This can be detected as it turns universal indicator paper blue.

6 What would you see if sodium hydroxide solution was added to a solution of:
  a Cu^{2+} ions
  b Fe^{2+} ions.

7 Using as few words as possible, describe the test for:
  a Al^{3+} ions
  b NH_4^+ ions.

P

3 Name the precipitate formed when sodium hydroxide reacts with:
  a aluminium nitrate solution
  b calcium nitrate solution.

4 Which ions could be present if the precipitate formed with sodium hydroxide is:
  a brown
  b white?

P

5 Write a balanced formula equation for the reaction, and state the colour of the precipitate, when sodium hydroxide solution is added to:
  a calcium chloride solution, $CaCl_2(aq)$
  b iron(III) nitrate solution, $Fe(NO_3)_3(aq)$

P

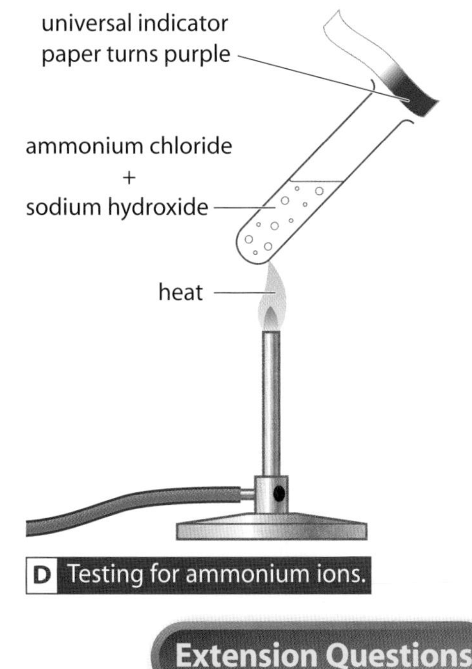

D Testing for ammonium ions.

Summary Exercise

Higher Questions

Extension Questions

4. Detecting acids

By the end of these two pages you should be able to:

- describe the tests for hydrogen (H^+) ions using acid/base indicators
- describe and explain the typical reactions of acids, using balanced formulae equations.

Gases such as sulphur dioxide and nitrogen dioxide, released by burning fossil fuels, make our rainwater more acidic than normal. This acid rain is a serious problem, as it damages metal structures, buildings and living things. Environmental chemists monitor the changes to our environment like the acidity of our rainwater.

1 **a** What are acid indicators?
  b Name two examples of indicators.

2 What causes acid rain?

3 What ion is in all acids?

0 1 2 3 4 5 6 7 8 9 10 11 12 13 14

more acidic — more alkaline

neutral

A The **pH scale** and the colours of universal indicator.

Water supplies in Britain are often acidic. Acidic solutions contain excess H^+ (hydrogen) ions. The H^+ ion can be detected using special **indicators** which change colour in **acids**. For example **litmus** and **universal indicator** both turn red in acids.

Acids can also be detected by two of their reactions that involve **effervescence** (fizzing). All acids react with fairly reactive metals such as magnesium, zinc and iron. For example, when magnesium reacts with hydrochloric acid the metal appears to dissolve and bubbles of gas are produced. The gas is hydrogen, which explodes with a 'pop' when lit. The reaction also forms a **salt** called magnesium chloride. As this salt is white and soluble it forms a colourless solution. The equations for this reaction are shown below.

$$2HCl + Mg \rightarrow MgCl_2 + H_2$$

hydrochloric acid + magnesium → magnesium chloride + hydrogen

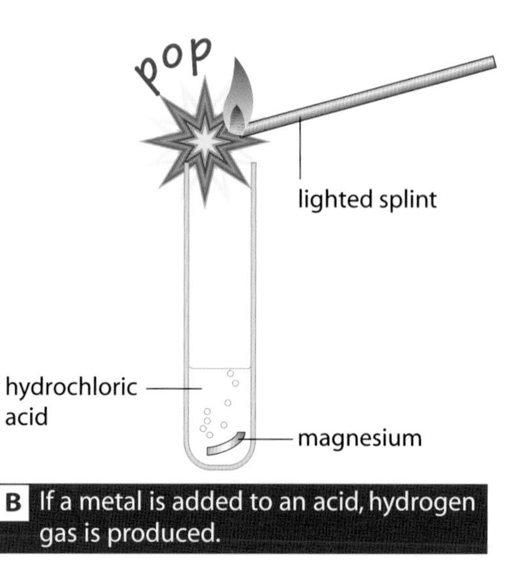

B If a metal is added to an acid, hydrogen gas is produced.

Different acids reacting with different metals produce different types of salt.

$$2HNO_3 + Zn \rightarrow Zn(NO_3)_2 + H_2$$

nitric acid + zinc → zinc nitrate + hydrogen

$$H_2SO_4 + Fe \rightarrow FeSO_4 + H_2$$

sulphuric acid + iron → iron sulphate + hydrogen

Acids also react with metal carbonates. For example, hydrochloric acid reacts with calcium carbonate producing bubbles of a gas, water and a salt. The gas can be identified as carbon dioxide because it turns **limewater** milky. The salt formed in this reaction is calcium chloride. As this salt is white and soluble it forms a clear solution.

$$2HCl + CaCO_3 \rightarrow CaCl_2 + H_2O + CO_2$$

hydrochloric acid + calcium carbonate → calcium chloride + water + carbon dioxide

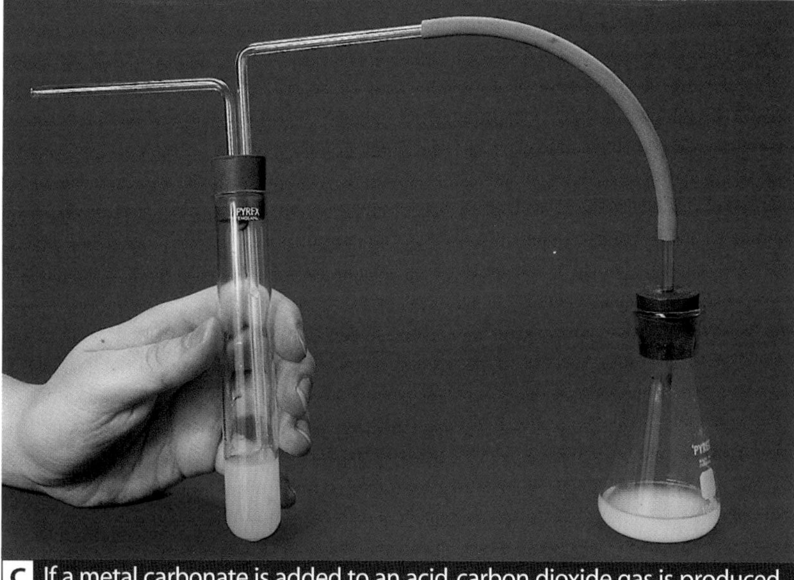

C If a metal carbonate is added to an acid, carbon dioxide gas is produced.

4 Images B and C show reactions of acids.
 a What do these reactions have in common?
 b What is the difference between these reactions?

5 Write a word and balanced chemical equation for the reaction between sulphuric acid and magnesium (Mg).

6 Write a word and balanced chemical equation for the reaction between sulphuric acid and copper(II) carbonate ($CuCO_3$).

7 Complete these general equations for the reactions of acids:
 a acid + metals → _____ + _____
 b acid + metal carbonates → _____ + _____ + _____

8 Describe three ways of testing for the presence of H^+ ions.

hydrochloric acid	HCl
sulphuric acid	H_2SO_4
nitric acid	HNO_3
phosphoric acid	H_3PO_4

D Formulae of common acids.

Summary Exercise

Higher Questions

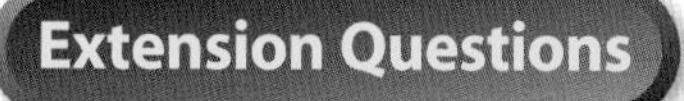

Extension Questions

5. Looking for anions

By the end of these two pages you should be able to:

- describe the tests for hydroxide (OH^-), carbonate (CO_3^{2-}), sulphite (SO_3^{2-}), sulphate (SO_4^{2-}), chloride (Cl^-), bromide (Br^-) and iodide (I^-) ions
- explain how precipitation reactions can form the basis of some tests for ions.

A Using indicators to test for hydroxide ions.

Once you have found the cations in a solution, you will need to carry out a series of tests for the anions. Litmus or universal indicator paper is a simple preliminary test. If the indicator turns blue/purple then hydroxide ions (OH^-) are present.

If hydroxide ions aren't present, try adding some hydrochloric acid. Bubbles of gas will be produced if carbonate ions (CO_3^{2-}) or sulphite ions (SO_3^{2-}) are present.

$$2HCl(aq) + Na_2CO_3(aq) \rightarrow 2NaCl(aq) + CO_2(g) + H_2O(l)$$

hydrochloric acid + sodium carbonate → sodium chloride + carbon dioxide + water

$$2HCl(aq) + Na_2SO_3(aq) \rightarrow 2NaCl(aq) + SO_2(g) + H_2O(l)$$

hydrochloric acid + sodium sulphite → sodium chloride + sulphur dioxide + water

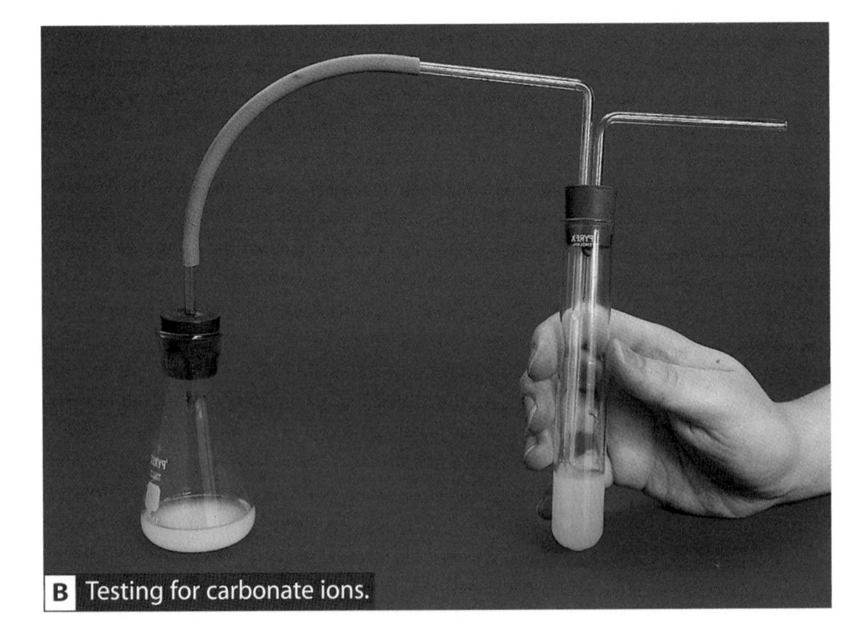

B Testing for carbonate ions.

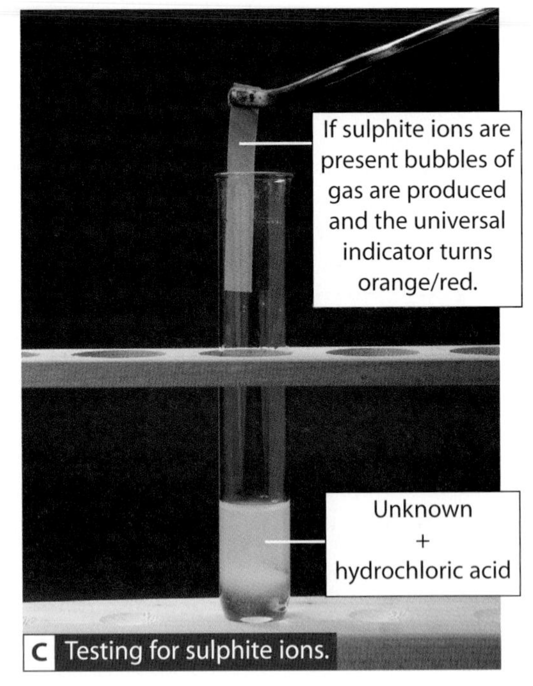

C Testing for sulphite ions.

You can identify which gas is formed, and the ion present, using the following tests.

- If the gas turns limewater milky, the gas is carbon dioxide, and a carbonate was present.
- If the gas is acidic and has a choking smell, the gas is sulphur dioxide, and a sulphite was present.

1 How do you test for hydroxide ions?

2 Bubbles of gas formed when hydrochloric acid was added to a white solid.

  a Which two ions could be present?

  b What would you do to find out which ion was present?

The test for sulphate ions (SO_4^{2-}) uses a precipitation reaction. If barium chloride solution is added to a sulphate, a white precipitate of barium sulphate forms.

$$Na_2SO_4(aq) + BaCl_2(aq) \rightarrow 2NaCl(aq) + BaSO_4(s)$$

sodium sulphate + barium chloride → sodium chloride + barium sulphate
WHITE

Carbonates and sulphites also produce a precipitate with barium chloride, so to be sure which ion is present in the sample you can add some hydrochloric acid. This has no effect on sulphates but dissolves the white precipitate of carbonates or sulphites.

D Testing for sulphate ions.

3 A white solid is either potassium sulphate or potassium sulphite. What could you do to identify the compound present?

4 What solid is formed when potassium sulphite is added to barium chloride solution?

The **halide ions**, chloride (Cl^-), bromide (Br^-) and iodide (I^-), can be identified using silver nitrate solution acidified with some dilute nitric acid. A different coloured precipitate of the silver halide forms in each case.

$$AgNO_3(aq) + NaCl(aq) \rightarrow AgCl(s) + NaNO_3(aq)$$

silver nitrate + sodium chloride → silver chloride + sodium nitrate
WHITE

F Testing for halide ions.

Halide ion	Precipitate with acidified silver nitrate solution
chloride (Cl^-)	white
bromide (Br^-)	cream
iodide (I^-)	yellow

E Testing for halide ions.

The dilute nitric acid is needed in this test to remove any carbonate, sulphite or hydroxide ions present as they also produce a precipitate with silver nitrate.

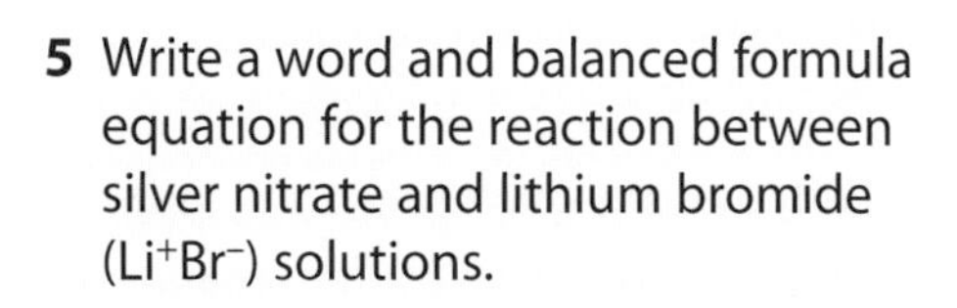

5 Write a word and balanced formula equation for the reaction between silver nitrate and lithium bromide (Li^+Br^-) solutions.

6 A solid could be copper sulphate or copper bromide. Describe the tests you could carry out to identify the solid.

Summary Exercise

Higher Questions

Extension Questions

6. Analysis in action

By the end of these two pages you should be able to:

- explain that ionic substances are identified by identifying each type of ion they contain.

A Forensic chemists must be very careful in their analysis.

Analytical chemists can be involved in testing a wide variety of substances. They could be asked to find the purity of drinking water or the source of pollutants in a river. They might have to test a consumer product, like a tinned food or a drain cleaner, to see if the contents label is correct. **Forensic chemists** working as scenes of crime officers can be asked to prove guilt or innocence by analysing materials found at a crime scene or in a suspect's home.

Have you ever wondered?

How does a forensic scientist work?

1 Describe three things that analytical chemists could do as part of their job.

Whatever the material, or the reasons for the analysis, the skills needed are the same. Analytical chemists must produce reliable results and this requires thorough planning, careful practical work, precise measurements and logical thinking.

For example, the household products in diagram B contain different ionic compounds. Imagine you needed to identify the ionic compound present in each of the products. Assuming they are all pure substances, where would you start your analysis? You will need to find at least two tests, from previous lessons, as each ionic compound contains two ions.

Test 1

Your first investigation would probably be to carry out flame tests. The colour a compound gives to a flame is a simple test for some common cations. The results are shown in table C.

Household chemical	Flame colour
indigestion tablets	yellow
fertiliser	none
moss killer	none
steriliser	lilac

C Flame tests.

B Some household chemicals.

2 What does test 1 tell you about the four chemicals?

Test 2

You could then add sodium hydroxide solution to a solution of each of the chemicals and look for precipitates or gases being produced. The results are shown in photograph D. Only the moss killer forms a precipitate, and it is pale green. The fertiliser didn't form a precipitate but it did produce a smelly alkaline gas on heating.

D Effect of sodium hydroxide solution.

Test 3

The next step would be to test for the anions present. You could start by adding some dilute hydrochloric acid and test any gas formed. The results of this test are given in table E.

Household chemical	Effect of adding dilute hydrochloric acid	Test on gas formed
indigestion tablets	bubbles of gas given off	limewater turns milky
fertiliser	no effect	–
moss killer	no effect	–
steriliser	bubbles of gas given off	acidic gas with a choking smell

E Testing for anions.

3 What does test 2 tell you about the four chemicals?

4 Look at table E. Which anions are present?

Test 4

For the final test the fertiliser and moss killer, which had given a negative result with hydrochloric acid, could be tested by the addition of barium chloride solution and acidified silver nitrate solution. See photograph F for the results.

These final tests should make it possible to identify all four chemicals. How did you do? Do you think you could be an analytical chemist?

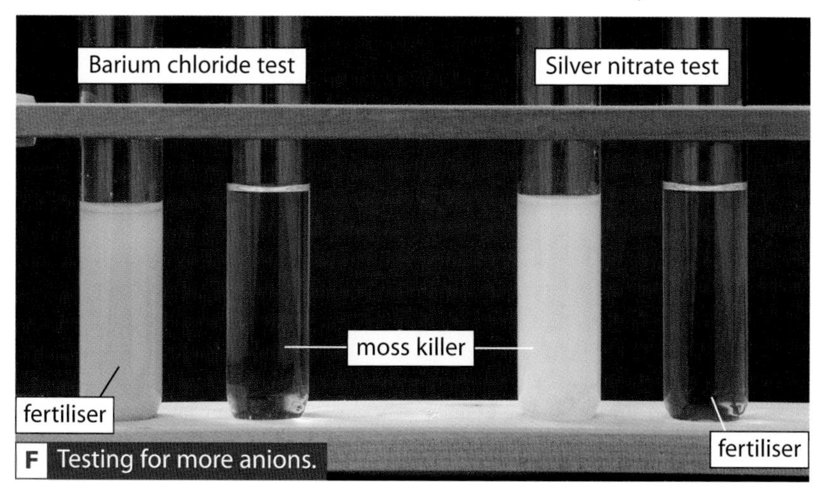

F Testing for more anions.

5 Look again at table E. What does it tell you about fertiliser and moss killer?

6 What does test 4 tell you about the moss killer and fertiliser?

7 Give the chemical name for each of the four household chemicals described above.

Higher Questions

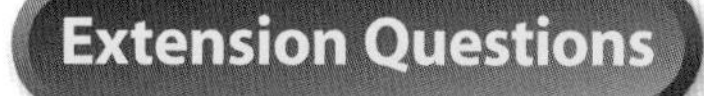

7. Explaining ion tests with equations

By the end of these two pages you should be able to:

- explain the importance of using the correct formulae for elements and compounds
- describe chemical reactions using balanced formulae equations
- write ionic equations for the reactions of acids and the tests for other ions.

To describe the simplest chemical reaction, we need to be able to write chemical **formulae** for the substances involved.

The formula of an ionic compound can be worked out using the charges on the ions involved.

Cations (positive ions)	Ion formula (negative ions)	Anions	Ion formula
sodium	Na^{+}	carbonate	CO_3^{2-}
calcium	Ca^{2+}	sulphite	SO_3^{2-}
ammonium	NH_4^{+}	sulphate	SO_4^{2-}
aluminium	Al^{3+}	chloride	Cl^{-}
calcium	Ca^{2+}	bromide	Br^{-}
iron(II)	Fe^{2+}	iodide	I^{-}
iron(III)	Fe^{3+}	hydroxide	OH^{-}

A Examples of ion formulae and charges.

This cross-over method involves the following steps.
Step 1: write the ions' symbols and charges.
Step 2: cancel down charge numbers if possible.
Step 3: cross over these numbers.
Step 4: tidy up the formula, put brackets where more than one ion exists and don't write 1s.

1 Use table A to work out the following formulae (show ion charges):
a sodium chloride
b calcium bromide
c iron (II) sulphate
d aluminium hydroxide.

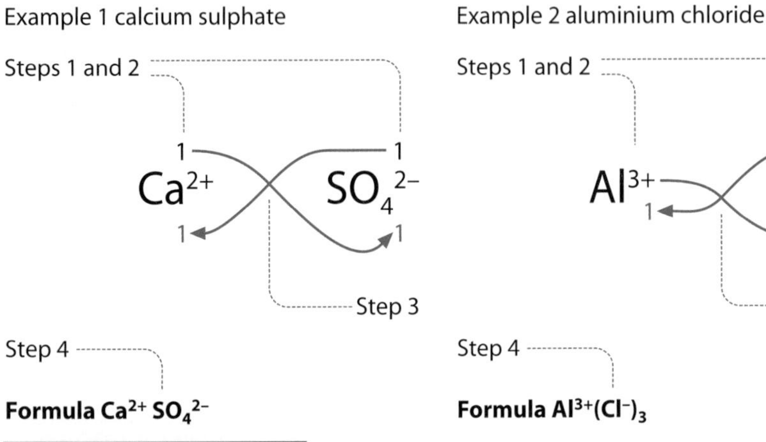

Formula $Ca^{2+}SO_4^{2-}$

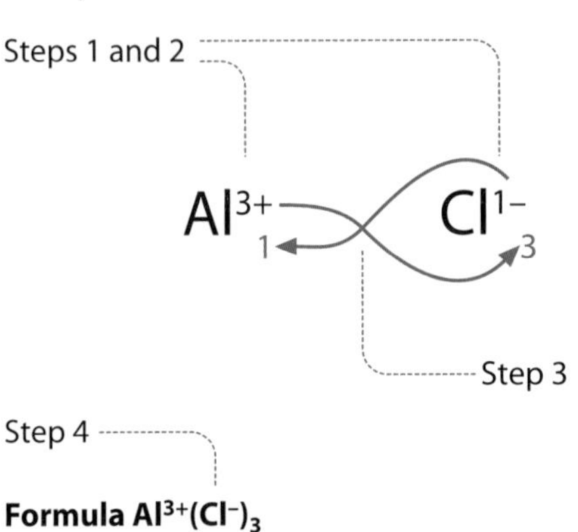

Formula $Al^{3+}(Cl^{-})_3$

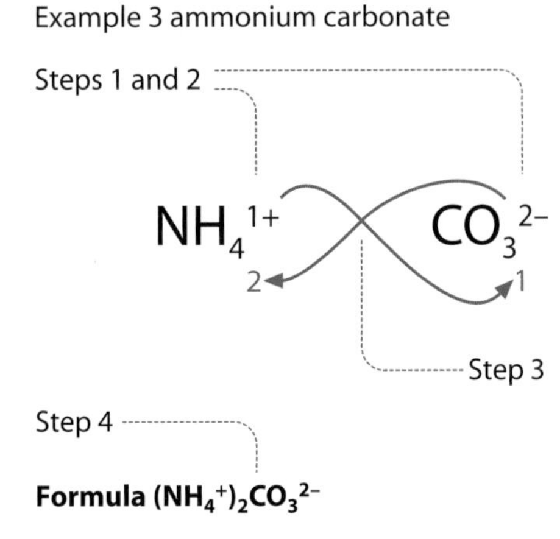

Formula $(NH_4^{+})_2CO_3^{2-}$

B The cross-over method.

To describe what happens during a chemical reaction, we can use these formulae to write different types of formulae equations.

Consider the balanced formula equation for the reaction between sodium carbonate and sulphuric acid.

$$H_2SO_4 + Na_2CO_3 \rightarrow Na_2SO_4 + H_2O + CO_2$$

This could be written showing the ion charges and the state symbols.

$$(H^+)_2\,SO_4^{2-}(aq) + (Na^+)_2CO_3^{2-} \rightarrow (Na^+)_2SO_4^{2-}(aq) + H_2O(l) + CO_2(g)$$

As the ions in solution can move about independently, they can be separated in the equation.

$$2H^+(aq) + SO_4^{2-}(aq) + 2Na^+(aq) + CO_3^{2-}(aq)$$
$$\rightarrow 2Na^+(aq) + SO_4^{2-}(aq) + H_2O(l) + CO_2(g)$$

The last equation shows that the $SO_4^{2-}(aq)$ and $Na^+(aq)$ ions are not changed by the reaction. These are called **spectator ions**. Removing them gives us an **ionic equation**, which only shows the changes taking place.

$$2H^+(aq) + CO_3^{2-}(aq) \rightarrow H_2O(l) + CO_2(g)$$

C Ions in solution are separated.

2 What is a spectator ion?

3 Look at the equation below.

$$H^+(aq) + Cl^-(aq) + Na^+(aq) + OH^-(aq) \rightarrow Na^+(aq) + Cl^-(aq) + H_2O(l)$$

a Which are the spectator ions?
b Write the ionic equation for this reaction.

4 What is an ionic equation?

5 What are the spectator ions in the barium chloride and sodium sulphate reaction?

The changes occurring in precipitation reactions are made more obvious by looking at their equations. Consider the addition of barium chloride solution to sodium sulphate solution.

Balanced equation:

$$(Na^+)_2SO_4^{2-}(aq) + Ba^{2+}(Cl^-)_2(aq) \rightarrow 2Na^+Cl^-(aq) + Ba^{2+}SO_4^{2-}(s)$$

Now separate the ions:

$$2Na^+(aq) + SO_4^{2-}(aq) + Ba^{2+}(aq) + 2Cl^-(aq)$$
$$\rightarrow 2Na^+(aq) + 2Cl^-(aq) + Ba^{2+}SO_4^{2-}(s)$$

Remove the spectator ions and produce the ionic equation:

$$SO_4^{2-}(aq) + Ba^{2+}(aq) \rightarrow Ba^{2+}SO_4^{2-}(s)$$

D Testing for sulphate ions involves a precipitation.

6 The equation below shows the reaction between sodium hydroxide and copper(II) sulphate.

$$Cu^{2+}(aq) + SO_4^{2-}(aq) + 2Na^+(aq) + 2OH^-(aq)$$
$$\rightarrow 2Na^+(aq) + SO_4^{2-}(aq) + Cu^{2+}(OH^-)_2(s)$$

a Which are the spectator ions in this reaction?
b Write the ionic equation for this reaction.

Summary Exercise

Higher Questions

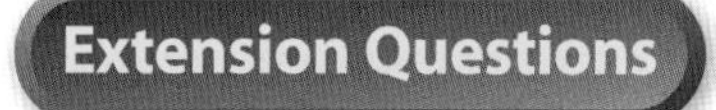

Extension Questions

8. How much is dissolved?

By the end of these two pages you should be able to:

- explain that the amount of a substance present in a solution can be found by experiments involving mass or concentration
- describe and explain how to find the mass of a substance dissolved in water by evaporating the water from a known mass of solution
- explain that different users require different levels of purity.

P

Sometimes we need to know how much of a substance is present. This is called quantitative analysis and involves techniques to find the **mass** of solid present or the **concentration** of **solute** in a solution.

For example, seawater contains sodium chloride and several other salts. The mass of all salts in 1 cubic decimetre of seawater can be found by the steps outlined below. (The symbol for cubic decimetres is dm^3.)

Conversion factors:

$1\ dm^3 = 1000\ cm^3 = 1000$ ml = 1 litre

1 Analysis can be qualitative or quantitative.

a What is the difference between them?

b What are the most common units of mass and concentration used in quantitative analysis?

V

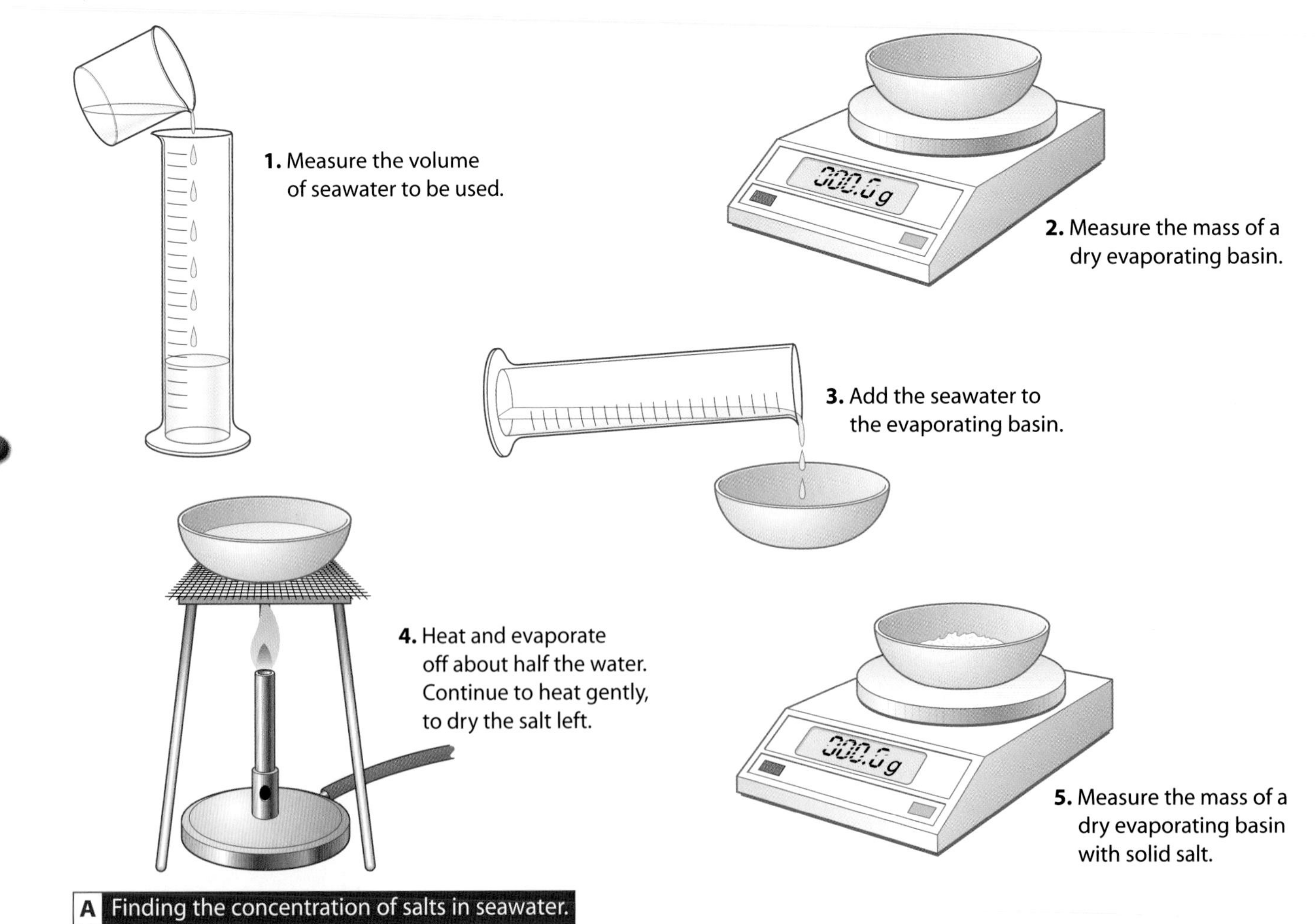

A Finding the concentration of salts in seawater.

Results

volume of seawater = 25 cm^3
mass of evaporating basin dry = 135.06 g
mass of evaporating basin + salt = 135.15 g

Calculation

mass of salt = 135.15 – 135.06 = 0.09 g
so 25 cm^3 seawater contains 0.09 g salt
concentration in 1000 cm^3 seawater
= 1000/25 × 0.09 = 3.6 g dm^{-3}

2 Seawater contains 3.6 g of salt per dm^{-3}. What mass of salt is in 500 dm^3?

3 What measurements must be taken to find the mass of solute in a certain volume of solution?

Different users of water require different levels of purity. For example, water for cleaning laboratory glassware can contain certain ions, but the water used to make up laboratory solutions has to be very pure. Our drinking water can be either hard or soft. Hard water contains more dissolved solids, for example calcium (Ca^{2+}) ion compounds. This could be good for your health, as calcium helps build strong bones and teeth. But there are disadvantages. Soap forms a scum in hard water and doesn't work properly. Hard water also produces limescale deposits in kettles and boilers, which makes them less efficient.

Chemists working in the laboratories of water authorities use similar techniques to measure the hardness of their water. They analyse the water by calculating the concentration of calcium carbonate in milligrams per cubic decimetre (mg dm^{-3}). These measurements are converted into the Clark's degrees scale (see table B) which is used for the settings on dishwashers and washing machines.

4 Apart from in laboratories, give another example of where the purity of water might be important.

5 What is hard water, and why is it sometimes a problem?

6 Describe in six or seven steps how you would find the concentration of a salt solution in g dm^{-3}.

Calcium carbonate (mg dm^{-3})	Hardness level	Clark's degrees
0 to 99.9	soft	0 to 6.9
100 to 199.9	moderately soft	7 to 13.9
200 to 299.9	moderately hard	14 to 20.9
300 to 399.9	hard	21 to 27.9
400 to 499.9	very hard	28 to 34.9

B Examples of ion formulae and charges.

7 a Exactly 250 cm^3 of hard water is evaporated in an evaporating basin. The dry basin weighed 186.12 g at the start and 186.19 g with solid salt. Calculate the concentration of solids in g dm^{-3} and mg dm^{-3}.

b Use table C to describe the hardness level of this water in words and in Clark's degrees.

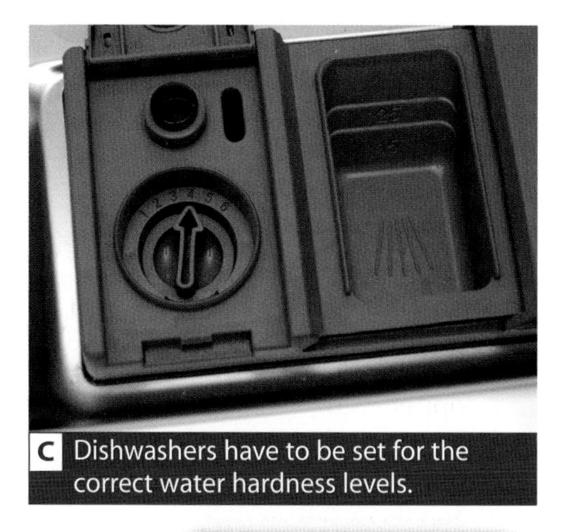

C Dishwashers have to be set for the correct water hardness levels.

Summary Exercise

Higher Questions

Extension Questions

9. Counting formulae

By the end of these two pages you should be able to:

- calculate the formula mass of elements and compounds
- explain that the amount of a substance can be measured in grams or moles.

When methane burns, one molecule of methane reacts with two molecules of oxygen to produce one molecule of carbon dioxide and two molecules of water.

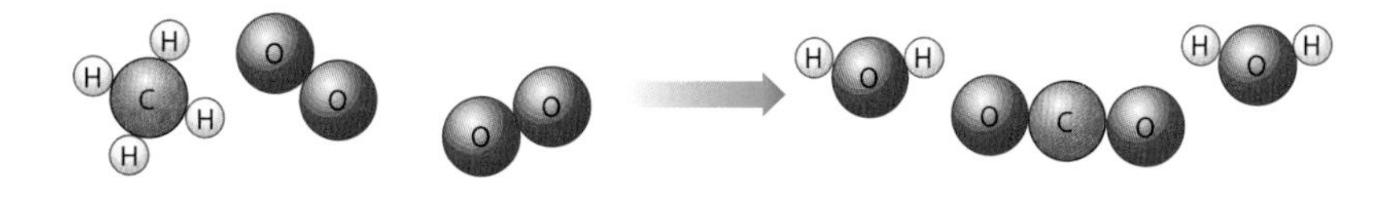

A The particles involved when burning methane are molecules.

The balanced formula equation shows how many molecules are involved.

$$CH_4 + 2O_2 \rightarrow CO_2 + 2H_2O$$

The formula used in equations represents the unit particle, which takes part in the reaction. Depending on the substance, these could be atoms, molecules or ion pairs. Using the correct formula is very important when calculating amounts in quantitative analysis. Some examples of different formulae are shown in table C.

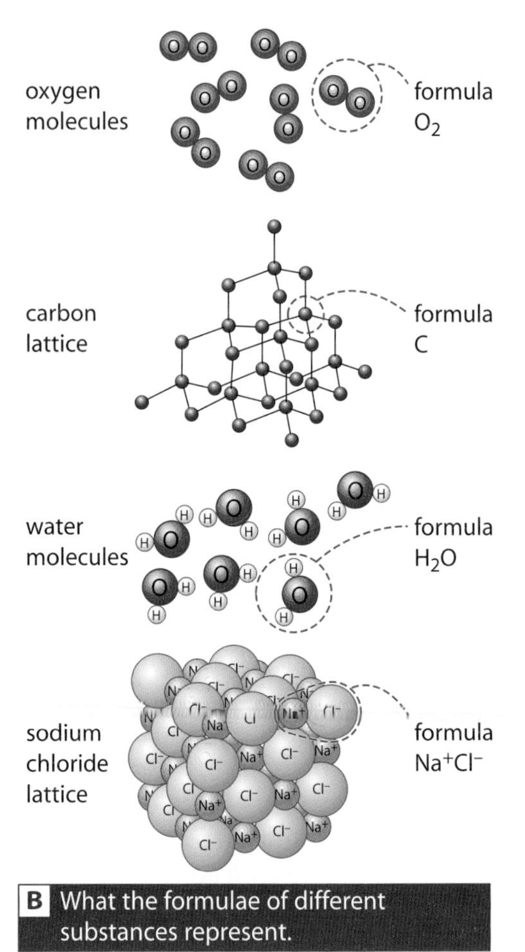

B What the formulae of different substances represent.

1 Copy and balance the following formulae equations:

a $C_3H_8 + O_2 \rightarrow CO_2 + H_2O$

b $Al + HCl \rightarrow AlCl_3 + H_2$

The formula of most **elements** is just their symbol, as this represents a single atom.	The formula of elements with **diatomic molecules** represents their molecules.	The formula of most **covalent compounds** represents a molecule of the substance.	The formulae of **ionic compounds** represent the ratio of the ion pairs present.
iron – Fe helium – He sulphur – S gold – Au copper – Cu	iodine – I_2 bromine – Br_2 chlorine – Cl_2 fluorine – F_2 oxygen – O_2 nitrogen – N_2 hydrogen – H_2	water – H_2O propane – C_3H_8 glucose – $C_6H_{12}O_6$ ammonia – NH_3 carbon dioxide – CO_2	aluminium oxide – $(Al^{3+})_2(O^{2-})_3$ copper sulphate – $Cu^{2+}SO_4^{2-}$ potassium bromide – $Ca^{2+}(Br^-)_2$

C Examples of formulae.

2 What is a diatomic molecule?

3 Explain why there are no molecular formulae for ionic compounds.

The relative masses of atoms are compared using **relative atomic masses** (RAM), A_r. These tell us approximately how much heavier one atom of the element is compared to a hydrogen atom. For example helium is roughly four times heavier than hydrogen. Relative atomic masses are shown on the Periodic Table or in separate tables of A_r values.

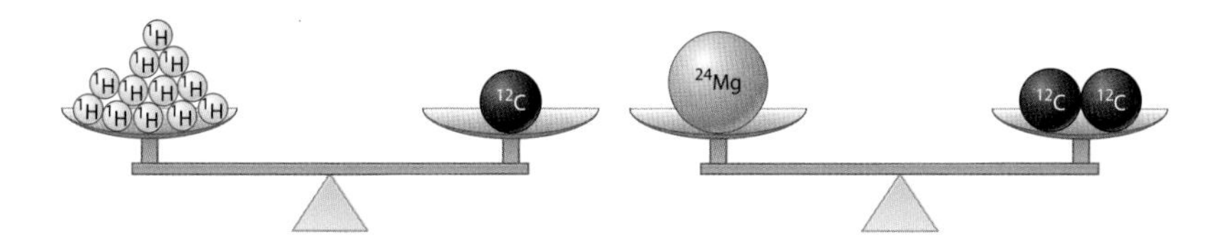

D Relative atomic masses.

4 a Which element in table E has atoms ten times heavier than a helium atom?

b How much heavier are magnesium atoms than carbon atoms?

The **relative formula mass** (also known as formula mass), M_r of an element or compound is found by adding together the relative atomic masses of all the atoms in the formula.

As relative atomic masses are just a comparison of masses, analytical chemists convert them into **moles**. A mole of a substance is equal to its relative formula mass in grams. For example, one mole of sodium carbonate weighs 106 g.

5 What is the formula mass of:

a $FeSO_4$ **b** $Mg(NO_3)_2$?

6 What is the mass of 1 mole of:

a CO_2 **b** $(NH_4)_2CO_3$?

7 Which of the following has the largest formula mass?

A HCl **B** HNO_3 **C** H_2SO_4 **D** H_2CO_3

8 What is the difference between the formula mass of a compound and one mole of a compound?

Element	Relative atomic mass
hydrogen	1
helium	4
carbon	12
nitrogen	14
oxygen	16
sodium	23
magnesium	24
aluminium	27
sulphur	32
chlorine	35.5
calcium	40
iron	56
lead	207

E Some common relative atomic masses.

Worked example

What is the relative formula mass of sodium carbonate?

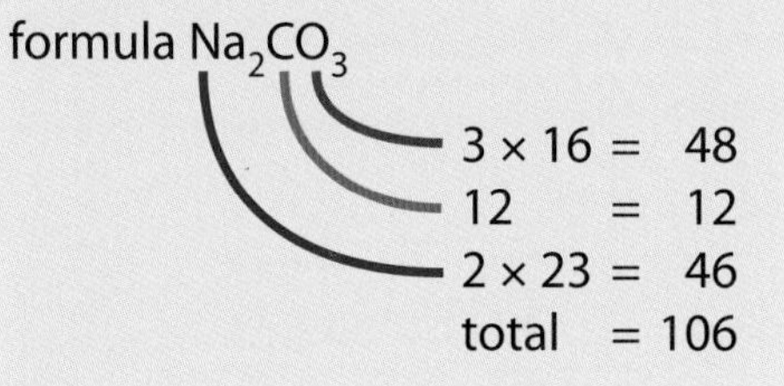

The relative formula mass = 106

Worked example

What is the mass of 1 mole of sodium hydroxide?

formula NaOH

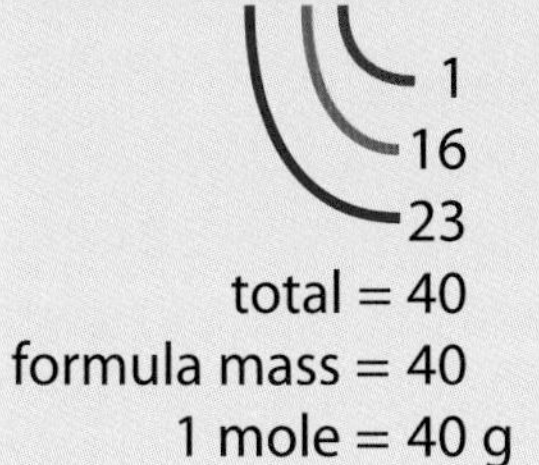

total = 40

formula mass = 40

1 mole = 40 g

Summary Exercise

Higher Questions

10. Measuring in moles

By the end of these two pages you should be able to:

- explain that the amount of a substance can be measured in grams, number of particles or moles
- describe how to convert masses into moles of particles
- describe how to convert moles into masses.

You have to use suitable units when you buy something in a shop. For example you might buy a metre of cloth, a kilogram of sugar, a dozen eggs or even a ream of paper.

The most appropriate unit for chemists is the mole, and you need to be able to convert moles to masses and masses to moles. The equation and memory triangle in diagram A can be used for these calculations.

1 Name four different units that can be used to measure amounts.

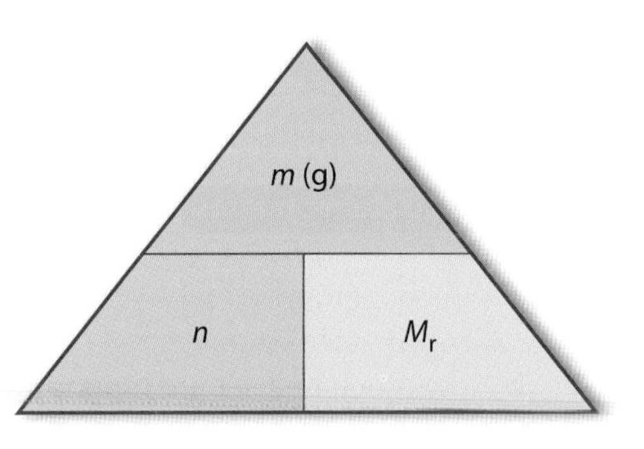

$m = n \times M_r$

m = mass of substance (g)
n = number of moles (mol)
M_r = formula mass (mass of 1 mole)

A Mass and moles memory triangle.

Worked example

What is the mass of 2.5 moles of iron oxide?
formula of iron oxide = Fe_2O_3

$16 \times 3 = 48$
$56 \times 2 = 112$
Total $= 160$

formula mass = 160
1 mole = 160 g
$m = n \times M_r$
$m = 2.5 \times 160 = 400$ g
2.5 moles Fe_2O_3 = 400 g

2 Find the mass of each of the following:
a 8 moles of sodium oxide (Na_2O)
b 0.4 moles of sulphur atoms(s)
c 5 moles of sodium carbonate (Na_2CO_3)
d 3.5 moles of water molecules (H_2O)

Worked example

How many moles of sulphuric acid are in 4.9 g of the acid?
formula of sulphuric acid = H_2SO_4

$4 \times 16 = 64$
$1 \times 32 = 32$
$2 \times 1 = 2$
Total $= 98$

formula mass = 98
1 mole = 98 g
$n = m/M_r$
$n = 4.9/98$
4.9 g of sulphuric acid = 0.05 moles

3 Find the the number of moles in each of the following:
a 96 g of oxygen (O_2)
b 60 g of carbon (C)
c 7.4 g of magnesium nitrate ($Mg(NO_3)_2$)
d 24.5 g of sulphuric acid (H_2SO_4).

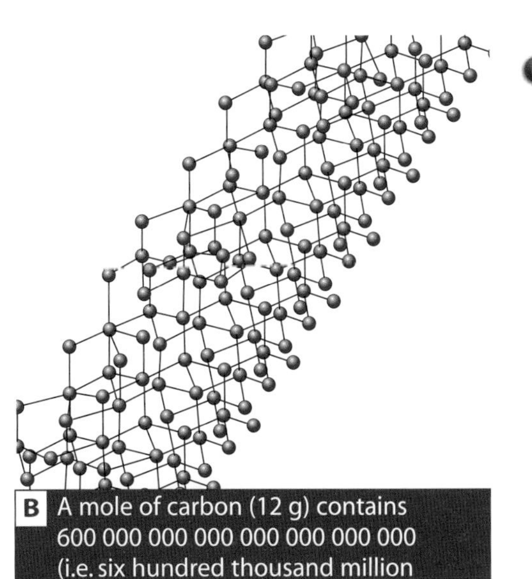

B A mole of carbon (12 g) contains 600 000 000 000 000 000 000 000 (i.e. six hundred thousand million million million atoms.)

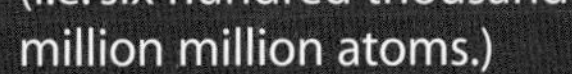

A mole of a substance is much more than just a mass in grams. It is also a fixed number of particles. This number is called the **Avogadro constant** and is equal to 6×10^{23} particles per mole (mol^{-1}). This is a very, very large number, but we can use it to carry out calculations involving moles, mass and numbers of particles.

The particles that make up a mole of a substance depend on its formula and can be described as 'formula units'. In calculations it is important to understand what the formula units represent as they may be atoms, molecules or ion pairs. Formula units are what the formulae represent. Some examples of formula units are shown in diagram C.

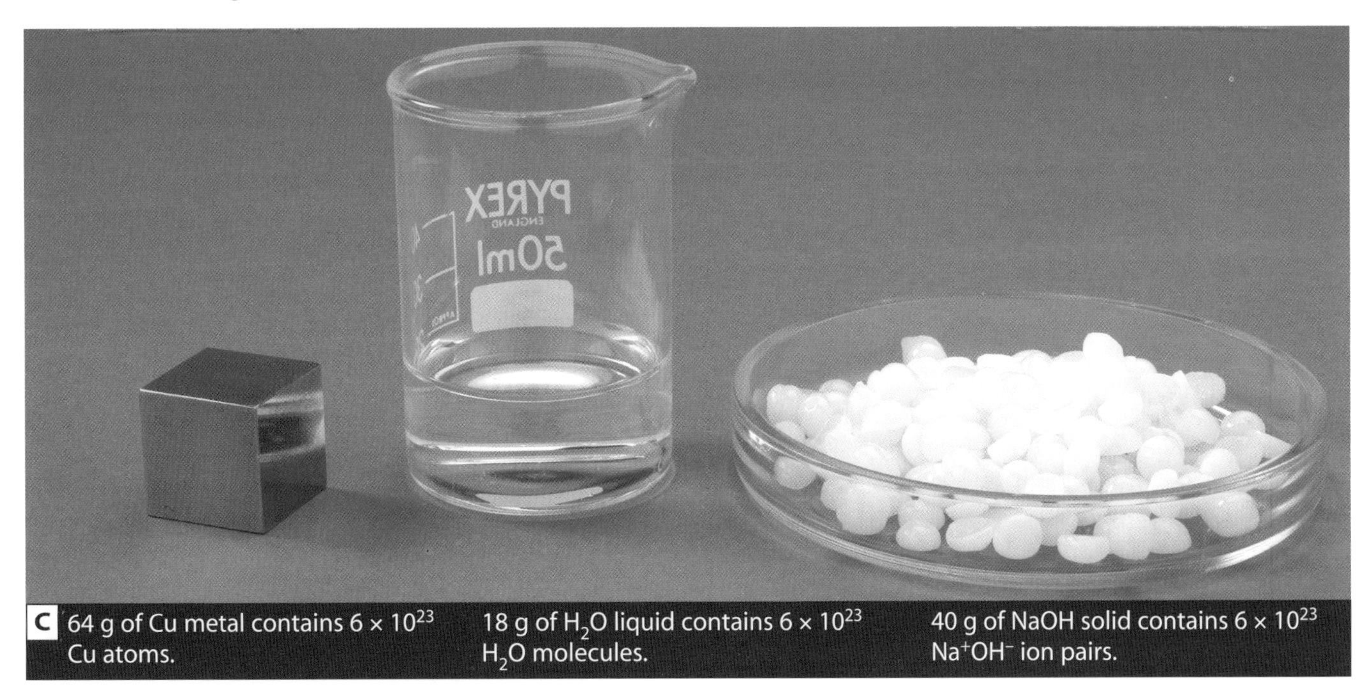

C 64 g of Cu metal contains 6×10^{23} Cu atoms. 18 g of H_2O liquid contains 6×10^{23} H_2O molecules. 40 g of NaOH solid contains 6×10^{23} Na^+OH^- ion pairs.

4 What is the same about 1 mole of any substance?

5 Which of the following contain the greatest number of atoms?
A 16 g of helium **B** 24 g of carbon
C 24 g of magnesium **D** 100 g of calcium

6 **a** What is the mass of 3.5 moles of calcium hydroxide ($Ca(OH)_2$)?
b How many moles are there in 37.4 g of ammonia (NH_3)?

7 Why is the unit of a 'mole' so important in chemistry?

Summary Exercise

Higher Questions

Extension Questions

11. How much reacts?

By the end of these two pages you should be able to:

- explain that to produce required amounts of product, chemists must be able to calculate how much reactant to use
- calculate the masses of substances involved in reactions, given the relevant equation.

Acids can be **neutralised** by a **base** to form a salt and water. For example, sulphuric acid reacts with sodium carbonate to form the salt sodium sulphate, water and also carbon dioxide.

If you don't add enough of the **reactant** sodium carbonate, there will be some acid left, so the solution at the end will not be neutral. If you add too much sodium carbonate the end solution will be alkaline. So you need to calculate the amount of acid and base needed to react completely. To do this you need to know the amount of acid present and the balanced formula equation for the reaction. The calculation will involve the following steps.

A Sodium carbonate neutralises sulphuric acid.

Step 1: Write the balanced formula equation.
Step 2: Find the mole ratios of substances in the question.
Step 3: Change the units as required.
Step 4: Solve the problem.

Worked example

What mass of sodium carbonate is needed to exactly neutralise 2.45 g of sulphuric acid?

1 Balanced equation
$Na_2CO_3 + H_2SO_4 \rightarrow Na_2SO_4 + H_2O + CO_2$

2 Mole ratios
1 mole $H_2SO_4 \equiv$ 1 mole Na_2CO_3

3 Change units
98 g $H_2SO_4 \equiv$ 106 g Na_2CO_3

4 Solve problem
1 g $H_2SO_4 \equiv$ 106/98 g Na_2CO_3
2.45 g $\equiv 2.45 \times 106/98 = 2.65$ g Na_2CO_3
2.65 g of sodium carbonate is needed to neutralise the acid.

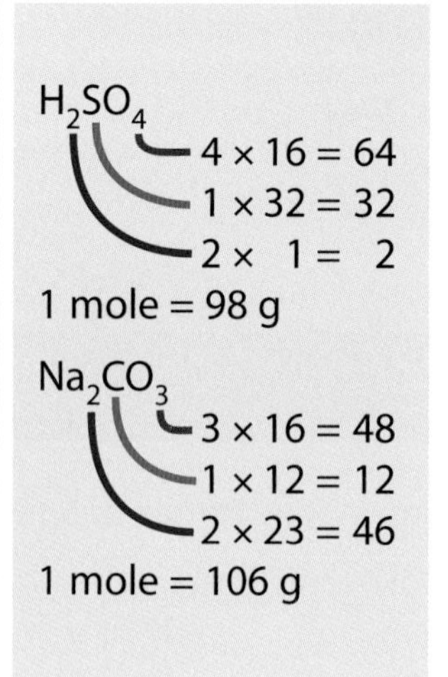

1 What mass of calcium carbonate is needed to neutralise 14.6 g of hydrochloric acid according to the equation:
$CaCO_3 + 2HCl \rightarrow CaCl_2 + H_2O + CO_2$

2 What mass of nitric acid is needed to react with 3.6 g of magnesium according to the equation:
$Mg + 2HNO_3 \rightarrow 2Mg(NO_3)_2 + H_2$

You can calculate the mass of solid formed in precipitation reactions in a similar way. For example, if sodium carbonate and silver nitrate solutions are added together, a precipitate of silver carbonate is formed. To calculate the mass of precipitate formed you need to know the mass of silver nitrate reacting and the balanced formula equation. The sodium carbonate must be added in excess, so that all the silver nitrate reacts.

Worked example

What mass of silver carbonate will precipitate when excess sodium carbonate solution is added to a solution containing 4.25 g of silver nitrate?

1 Balanced equation
$Na_2CO_3(aq) + 2AgNO_3(aq) \rightarrow Ag_2CO_3(s) + 2NaNO_3(aq)$

2 Mole ratios
2 mole $AgNO_3$ ≡ 1 mole Ag_2CO_3

3 Change units
340 g $AgNO_3$ ≡ 276 g Ag_2CO_3

4 Solve problem
1 g $AgNO_3$ ≡ 276/340 g Ag_2CO_3
4.25 g ≡ 4.25 × 276/340 = 3.45 g Ag_2CO_3
3.45 g of silver carbonate is precipitated.

$AgNO_3$
3 × 16 = 48
1 × 14 = 14
1 × 108 = 108
1 mole = 170 g

2 mole = 2 × 170 = 340 g

Ag_2CO_3
3 × 16 = 48
1 × 12 = 12
2 × 108 = 216
1 mole = 276 g

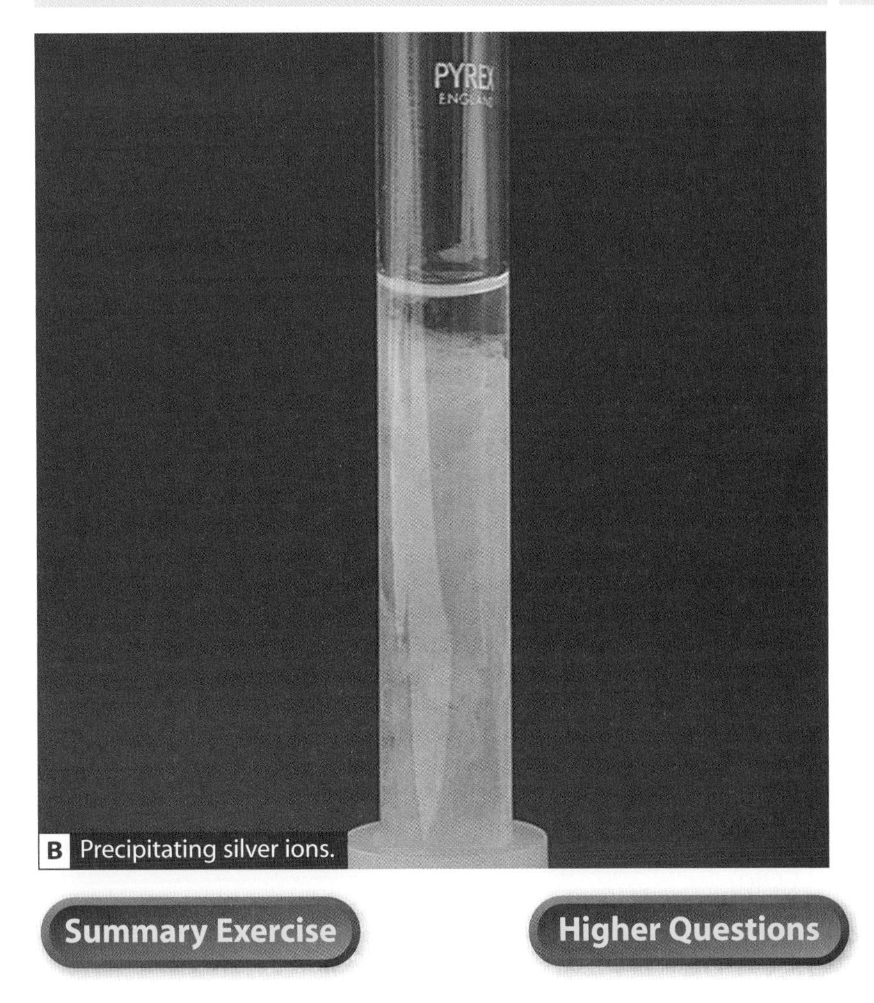

B Precipitating silver ions.

3 What mass of silver chloride will be formed when excess silver nitrate is added to a solution containing 11.7 g of sodium chloride? The equation is:
$AgNO_3 + NaCl \rightarrow NaNO_3 + AgCl$

4 What mass of magnesium sulphate can be formed by reacting 1.2 g of magnesium metal with excess sulphuric acid? The equation is:
$Mg + H_2SO_4 \rightarrow MgSO_4 + H_2$

5 When magnesium oxide reacts with hydrochloric acid a salt called magnesium chloride is formed. What do you need to know to calculate the mass of magnesium chloride formed in this reaction?

Summary Exercise
Higher Questions
Extension Questions
P

12. Avogadro's law and gas volumes

By the end of these two pages you should be able to:

- use Avogadro's law to calculate the volumes of gases involved in reactions, given the relevant equation.

Gases are sometimes difficult to measure and contain because their molecules move about freely in all directions. If a gas escapes it can be dangerous, as it will spread out quickly and fill the room. In addition although we can easily measure the volume of a gas, it is difficult to measure its mass.

In 1811 the Italian scientist Amedeo Avogadro proposed an idea that is now used in calculations involving gases. This idea is called **Avogadro's law** and it states that, 'at the same temperature and pressure, equal volumes of gases will contain equal numbers of gas particles.' This has now been found to be true for most gases.

A Collecting and measuring a gas.

1 Draw a labelled diagram to show how you would collect and measure the gas produced in the reaction of magnesium and hydrochloric acid.

2 What is Avogadro's law?

Worked example

A student wanted to investigate Avogadro's law and set up the apparatus shown below.

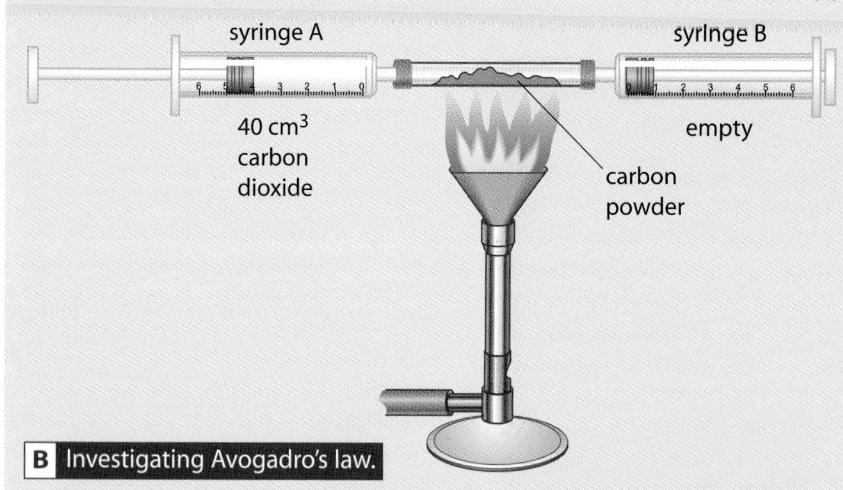

B Investigating Avogadro's law.

40 cm^3 of carbon dioxide gas in syringe A was passed back and forth over the heated carbon, until all the carbon dioxide was converted to carbon monoxide.

If the gas was allowed to cool to its original temperature, what would be the final volume of the gas?

The balanced equation tells us the ratio in molecules.
$CO_2(g) + C(s) \rightarrow 2CO(g)$
1 molecule of $CO_2(g) \rightarrow$ 2 molecules of $CO(g)$

By Avogadro's law, equal volumes contain equal numbers of molecules
1 volume of $CO_2(g) \rightarrow$ 2 volumes $CO(g)$

Substituting the volume of gas used in the question:
40 cm^3 of $CO_2(g) \rightarrow 2 \times 40$ cm^3= 80 cm^3 $CO(g)$

The final volume of carbon monoxide gas is 80 cm^3.

3 Pentane can burn according to the equation:

$C_5H_{12}(g) + 8O_2(g) \rightarrow 5CO_2(g) + 6H_2O(l)$

If 50 cm^3 of pentane is burned, what volume of:

a oxygen is needed?

b carbon dioxide is formed?

Gases obey Avogadro's Law because the distances between their molecules are very much bigger than in solids or liquids. This means the volume of a gas only depends on the distance between its molecules, not their size. Since the distance between all gas molecules is similar, equal volumes of gases contain the same number of molecules.

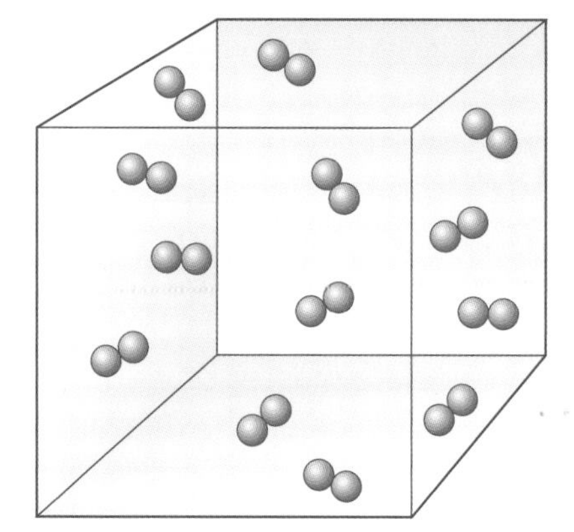

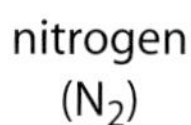

nitrogen (N_2)

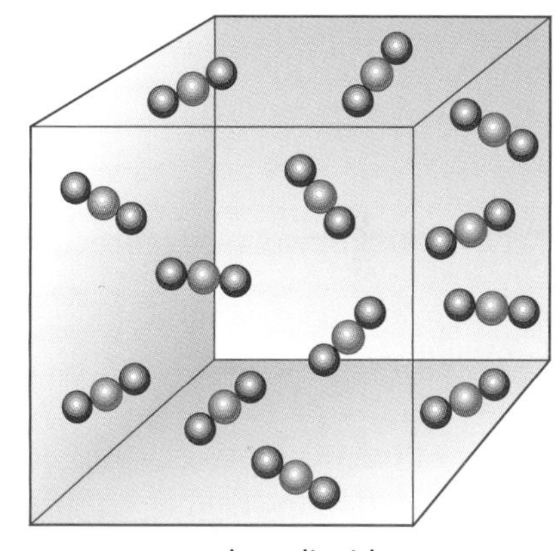

carbon dioxide (CO_2)

C The same volume of different gases contains the same number of gas molecules.

Worked example

If 60 cm^3 of ammonia gas is completely decomposed, what would be the total final volume of gases produced? (All measurements made at the same temperature and pressure.)

The balanced equation tells us the ratio in molecules.

$2NH_3(g) \rightarrow N_2(g) + 3H_2(g)$

The ratio in molecules is:

2 molecules $NH_3 \rightarrow$ 1 molecule N_2 + 3 molecules H_2

By Avogadro's law

2 volumes $NH_3 \rightarrow$ 1 volume N_2 + 3 volumes H_2

Substituting the volume of gas used in the question:

$60\ cm^3\ NH_3 \rightarrow \frac{1}{2} \times 60\ cm^3\ N_2 + \frac{3}{2} \times 60\ cm^3\ H_2$

$= 30\ cm^3\ N_2 + 90\ cm^3\ H_2$

The total final volume of gas is 120 cm^3.

4 Methane and steam react as shown below:

$CH_4(g) + H_2O(g) \rightarrow CO(g) + 3H_2(g)$

If 500 cm^3 of methane reacts what would be the total volume of products formed?

5 Explain why the same volumes of different gases contain the same number of molecules.

Summary Exercise

Higher Questions

13. Working with gases

By the end of these two pages you should be able to:

- calculate the volume of a given mass of gas given the molar volume at the appropriate temperature and pressure and *vice versa*
- calculate masses or volumes of substances involved in a reaction, given the relevant equation.

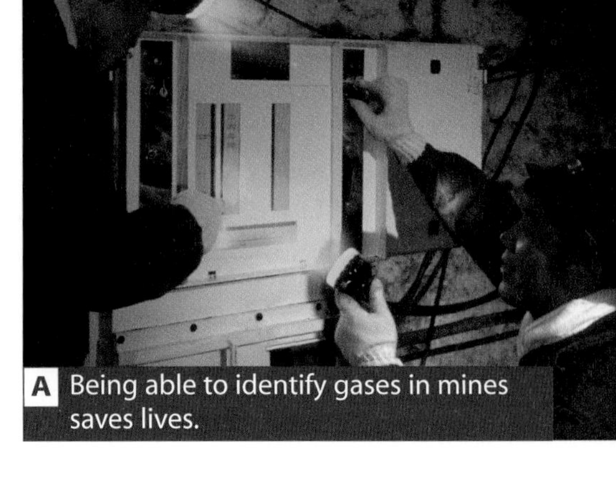

A Being able to identify gases in mines saves lives.

The breakdown of plant material produces a mixture of different gases including hydrogen, carbon monoxide, hydrogen sulphide and methane. This mixture of gases is produced in marshes and in coal mines, where it is called firedamp. These gases have been responsible for the deaths of many miners, as they are both poisonous and explosive. One of the most important jobs of mine safety officers is to test for these gases.

1 Write the formulae for three of the gases in firedamp.

2 Name one gas in firedamp that is poisonous and one which is explosive.

The relationship between moles and gas volumes is described in Avogadro's law. That is, equal volumes of gases contain equal numbers of gas molecules (measured at the same temperature and pressure). This means that equal numbers of gas molecules must take up the same volume. Since a mole of anything contains the same number of particles (6×10^{23}), then one mole of any gas must have the same volume. This is called the **molar volume** and its value, at a particular temperature and pressure, can be used in calculations involving moles, mass and volumes of gases. At normal room temperature and pressure the molar gas volume is about 24 dm^3. The equation and memory triangle in diagram B can be used for these calculations.

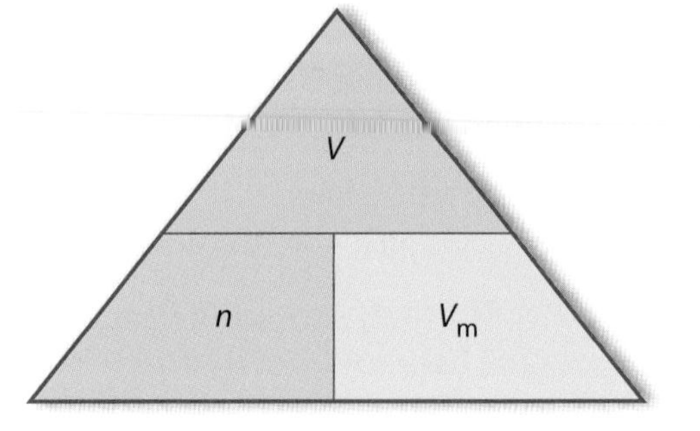

$n = V / V_m$

V = volume of gas (dm^3)

n = number of moles (mol)

V_m = molar volume

B Gas volumes memory triangle.

Worked example

The molar volume of gases is 24 dm^3. A student collected 200 cm^3 of carbon dioxide gas. What mass of the gas was collected?

First we need to find the number of moles in 200 cm^3 of carbon dioxide:

$n = V/V_m$

$n = 200/24000 = 0.00833$ moles

As 1 mole of CO_2 = 44 g

0.00833 moles = 0.00833×44

200 cm^3 CO_2 = 0.367 g

CO_2

$2 \times 16 = 32$

$1 \times 12 = 12$

1 mole = 44 g

3 If the molar volume is 24 dm^3:

a what is the mass of 1 dm^3 of oxygen (O_2)

b what is the volume of 10 g of helium (He)?

Worked example

To find the molar volume of hydrogen at room temperature the class set up the apparatus shown in diagram C.

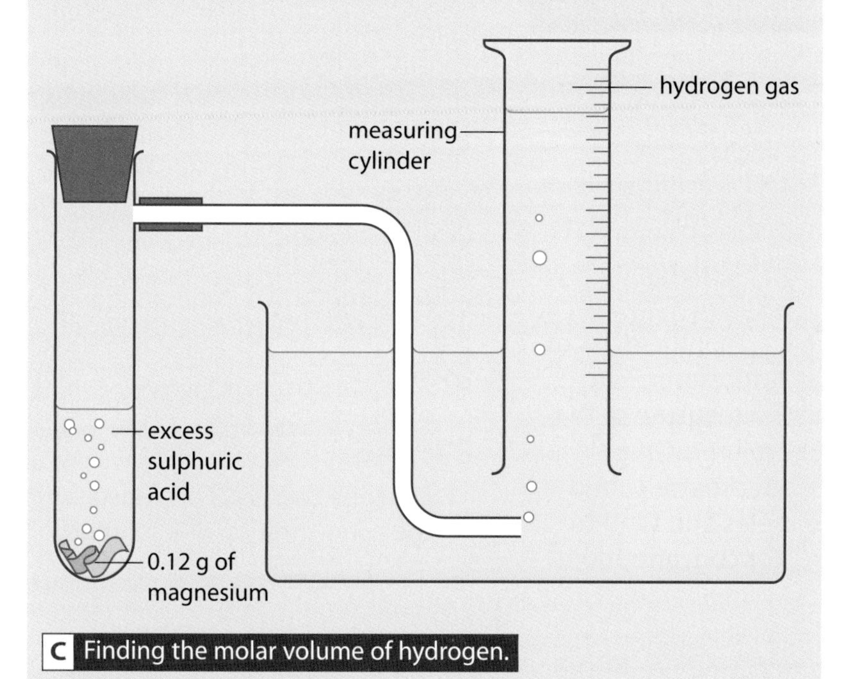

C Finding the molar volume of hydrogen.

Results

Mass of magnesium = 0.12 g
Volume of hydrogen = 118 cm^3

According to the balanced equation
$Mg + H_2SO_4 \rightarrow MgSO_4 + H_2$
1 mole Mg ≡ 1 mole H_2
1 mole Mg = 24 g
0.12 g Mg = 0.12/44 g = 0.005 moles
0.005 moles (0.12 g) Mg ≡ 0.005 moles H_2
0.005 moles of H_2 ≡ 118 cm^3 H_2
1 mole of H_2 ≡ 118/0.005
molar volume = 23 600 cm^3 (23.6 dm^3)

D A mole of any gas under normal conditions has a volume of about 24 dm^3.

4 What volume of oxygen would be produced if 1.36 g of hydrogen peroxide decomposes?
$2H_2O_2(l) \rightarrow 2H_2O(l) + O_2(g)$
(molar volume 24 dm^3)

5 Which of the following has the greatest volume?
A 18 g of $O_2(g)$ **B** 34 g of $SO_2(g)$
C 12 g of $CH_4(g)$ **D** 20 g of $N_2(g)$

P

Higher Questions

14. Looking at solutions

By the end of these two pages you should be able to:

- convert concentration in g dm^{-3} into mol dm^{-3}, and *vice versa*.

Have you ever wondered?

Why do we need to analyse substances?

Analytical chemists, working for the Trading Standards Department, check the contents of products to see if manufacturers' claims are true. On a contents label the mass of each solid is usually given in grams (g). If the substance is a solution, it is given as a concentration. This is a measure of the amount of a substance dissolved in a given volume. On a contents list it is usually given in grams per decimetre cubed (g dm^{-3}). However, chemists often work in moles and tend to use moles per decimetre cubed (mol dm^{-3}) for concentrations.

1 Describe two ways of measuring the concentration of a solution.

Have you ever wondered?

Why is it important to know that the label of contents on the packet is correct?

To change concentrations from one form to another you need to know the mass of one mole of the solute (symbol M_r).

Thus: g dm^{-3} = mol dm^{-3} × M_r

Worked example

The label on a bottle of 'Bleacho' claimed it contained at least 65 g dm^{-3} of sodium hypochlorite (NaOCl). Chemists in a trading standards laboratory found the concentration of sodium hypochlorite in 'Bleacho' to be 0.65 mol dm^{-3}. Was the label correct?

Formula NaOCl

1 × 35.5 = 35.5
1 × 16 = 16
1 × 23 = 23

1 mole = 74.5 g

Now g dm^{-3} = mol dm^{-3} × M_r
= 0.65 × 74.5
concentration = 48.43 g dm^{-3}
and so the label was wrong.

A Is this the strongest bleach?

For concentrations in grams per decimetre (g dm^{-3}).

$C = \frac{m}{V}$ where

C = concentration (g dm^{-3})
m = mass of solute (g)
V = volume of solution (dm^3)

For concentrations in moles per decimetre (mol dm^{-3}).

$C = \frac{n}{V}$ where

C = concentration (mol dm^{-3})
n = number of moles (mol)
V = volume of solution (dm^3)

Conversion factors:
1 dm^3 = 1000 cm^3 = 1000 ml = 1 litre

2 Normal laboratory hydrochloric acid has a concentration of 1 mol dm^{-3}. What is its concentration in g dm^{-3}?

3 A solution of hydrogen peroxide (H_2O_2) has a concentration of 1.7 g dm^{-3}. Convert this concentration into mol dm^{-3}.

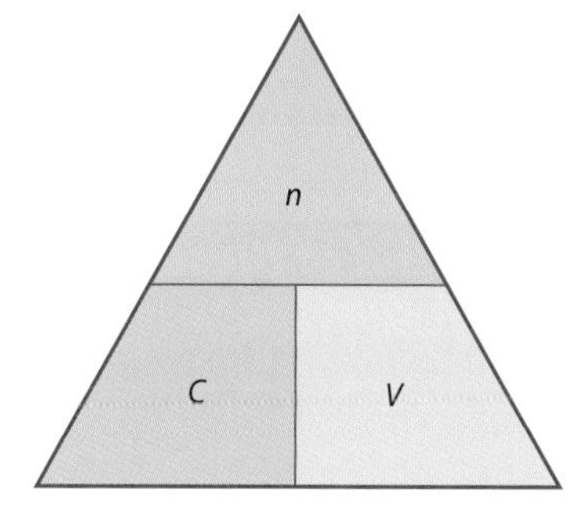

$C = n / V$

n = number of moles (mol)
C = concentration (mol dm^{-3})
V = volume (dm^3)

B Moles and concentration memory triangle.

Worked example

The contents label on a bottle of vinegar, chemical name ethanoic acid, claims that the concentration of acid is 38.5 g dm^{-3}. (1 mole of ethanoic acid, CH_3COOH is 60 g.)

a What mass of ethanoic acid is dissolved in 150 cm^3 of vinegar?

b What is the concentration of ethanoic acid in vinegar in mol dm^{-3}?

a $C = m/V$
$m = C \times V$
Substituting values into the question
$m = 38.5 \times 0.15$
$m = 5.78$ g ethanoic acid

b 1 mole CH_3COOH = 60 g
g dm^{-3} = mol $dm^{-3} \times M_r$
Substituting values in question
38.5 = mol $dm^{-3} \times 60$
mol dm^{-3} = 38.5/60
concentration CH_3COOH = 0.64 mol dm^{-3}

C The concentration of household chemicals can be measured in g dm^{-3}.

4 What is the concentration (in mol dm^{-3}) of a 250 cm^3 solution containing 1.6 g of sodium hydroxide?

5 Which of the following solutions of sodium chloride contains the greatest mass of solute?
A 200 cm^3 of 2 mol dm^{-3} **B** 1000 cm^3 of 0.4 mol dm^{-3}
C 100 cm^3 of 5 mol dm^{-3} **D** 400 cm^3 of 1 mol dm^{-3}

6 A solution is made by dissolving 12 g of sodium hydroxide (NaOH) in 200 cm^3 of solution. Calculate the concentration of this solution in two different standard units.

Summary Exercise | Higher Questions | Extension Questions

P

15. Titration and standard solutions

By the end of these two pages you should be able to:

- describe the procedure for carrying out simple acid–base titrations using burette, pipette and suitable indicators.

Finding the volumes of different solutions that react together and the concentrations of unknown solutions is called **volumetric analysis**. The experimental technique used is called a **titration**. This requires precise equipment, careful manipulation and accurate measurement. Volumetric analysis is an extremely accurate way of analysing solutions.

Many different acids and **alkalis** are used in everyday life. These solutions are often analysed using titrations. In these titrations a **pipette** is used to accurately measure a set volume of the alkali into a conical flask. Pipettes come in several sizes varying from 1 cm^3 to 50 cm^3 (accuracy ±0.05 cm^3.) An indicator is added to the alkali in the conical flask. The acid is placed in a **burette**. This can measure the exact volume of the acid solution added during the titration (accuracy ±0.1 cm^3.) The acid in the burette is slowly added to the alkali in the conical flask until the indicator changes colour. The volume of acid solution added is then noted by reading the scale on the burette.

A A student carrying out a titration.

1 What is volumetric analysis?

2 What are the following pieces of apparatus used for?
a burette
b pipette

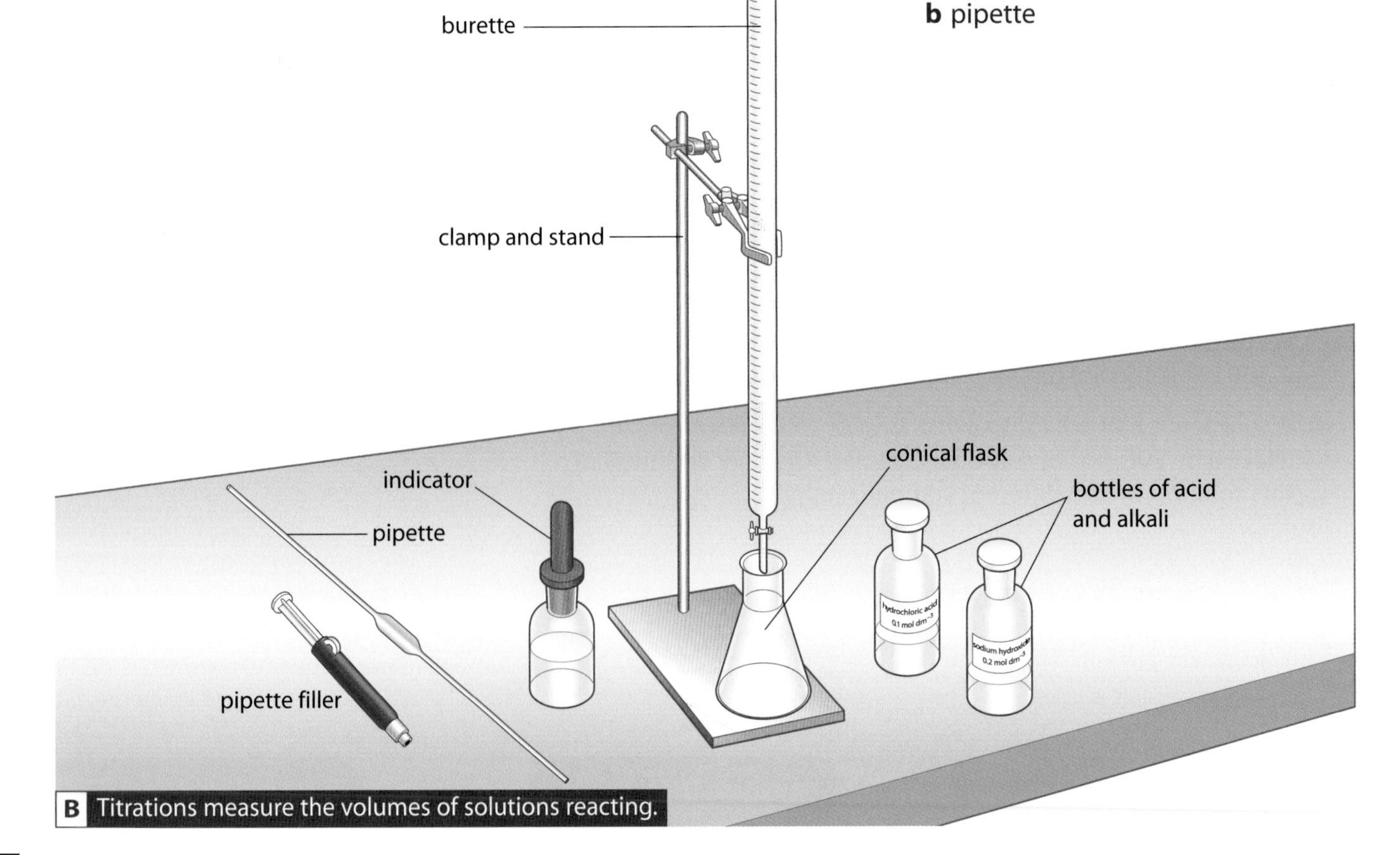

B Titrations measure the volumes of solutions reacting.

For accurate results in volumetric analysis we need to use solutions of an exact, known concentration. These solutions must react completely with our solution of unknown concentration, be stable on storage and have a sharp colour change with the indicator. Solutions with these properties are called **standard solutions.** It is an important task of technicians in all analytical laboratories to make up these solutions.

The steps involved in producing an accurate standard solution are as follows.

Step 1: Calculate the mass of solute required.
Step 2: Weigh out the exact mass of solute using a balance.
Step 3: Dissolve the solute in a small amount of water in a beaker.
Step 4: Transfer the solution to a **volumetric flask**.
Step 5: Wash beaker with distilled water and transfer 'washings' to the flask.
Step 6: Add distilled water to volumetric flask to make up to graduation mark.
Step 7: Stopper flask and invert two or three times to mix.

C A volumetric flask.

Worked example

Sodium carbonate (Na_2CO_3) makes an excellent standard solution for analysing acids. What mass of sodium carbonate is needed to make up 250 cm^3 of 0.2 mol dm^{-3} solution?

As $n = C \times V$

$n = 0.2 \times 0.25$ (The volume has been converted to dm^3.)

moles of Na_2CO_3 needed = 0.05 moles

Na_2CO_3
- $3 \times 16 = 48$
- $1 \times 12 = 12$
- $2 \times 23 = 46$

1 mole = 106 g

mass = moles × relative formula mass
mass of 0.05 moles $Na_2CO_3 = 0.05 \times 106$
mass of Na_2CO_3 needed = 5.3 g

Remember that to calculate concentrations in chemistry we use:
$n = C \times V$
where n = number of moles (mol)
C = concentration (mol dm^{-3})
V = volume of solution (dm^3)

D Concentrations of laboratory solutions are measured in mol dm^{-3}.

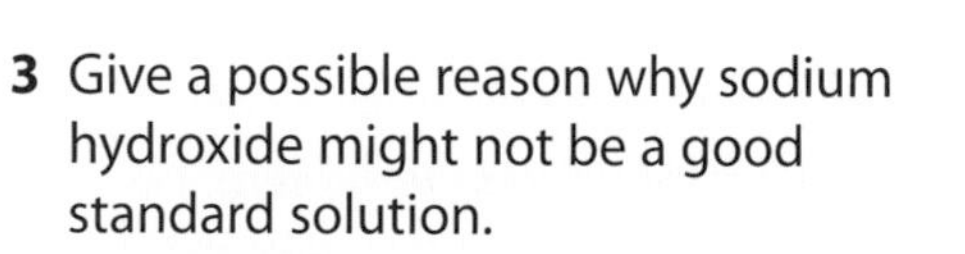

3 Give a possible reason why sodium hydroxide might not be a good standard solution.

4 What mass of sodium carbonate is needed to make 200 cm^3 of 0.05 mol dm^{-3} solution?

5 Write brief instructions to describe the steps involved in a titration. Include the following words in your answer: acid; alkali; burette; indicator and pipette.

Summary Exercise | **Higher Questions** | **Extension Questions**

16. Volumetric analysis

By the end of these two pages you should be able to:

- explain how the amount of a substance dissolved in a solution can be found by titration
- carry out simple calculations from the results of titration.

Finding the concentration of a solution is a routine task in chemistry laboratories throughout the world. Using titrations we can analyse pollutants in waste water, verify the accuracy of manufacturers' descriptions, monitor patients' health in hospitals and check the consistency of solutions used in industry.

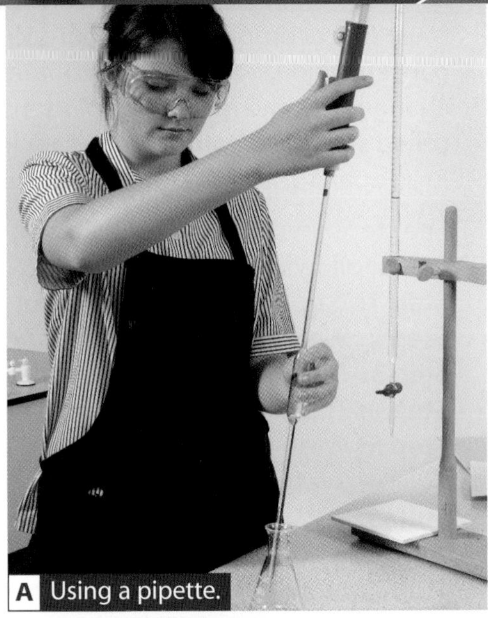

A Using a pipette.

Have you ever wondered?

How do we find out how much of a substance is present in a given sample?

1 Give two examples of uses of volumetric analysis.

The steps involved in using standard sodium carbonate solution to analyse hydrochloric acid are outlined below.

Step 1: Use a pipette to measure 25 cm^3 of standard sodium carbonate into a conical flask and add indicator (see diagram A).

Step 2: Fill a burette with the hydrochloric acid and note the level of the acid in the burette.

Step 3: Add the acid to the solution in the conical flask until the indicator changes colour. Note the new level of the acid in the burette (see diagram B).

Step 4: Repeat steps 1–3, recording the results for each titration carried out.

B Using a burette.

Table C shows how the results should be recorded.

Titration	1st level (cm^3)	2nd level (cm^3)	Volume of acid (cm^3)	Average volume of acid (cm^3)
1 (rough)	0.0	8.2	8.2	Average = (7.5 + 7.3)/2 = 7.4 cm^3 (Ignore first, rough, titration when averaging.)
2	8.2	15.7	7.5	
3	15.7	23.0	7.3	

C Titration results.

2 The volumes of acid added during a titration were:
1st (rough) = 27.2 cm^3; 2nd = 26.4 cm^3; 3rd = 26.2 cm^3.
What volume would you use in the calculation?

The unknown concentration of the hydrochloric acid can be calculated using the results of the titration and a balanced formula equation for the reaction.

Worked example

Results:

volume of sodium carbonate = 25 cm^3
concentration of sodium carbonate 0.2 mol dm^{-3}
volume of hydrochloric acid = 7.4 cm^3
concentration of hydrochloric acid = ? mol dm^{-3}

First write down the balanced equation for the reaction, as this tells us the mole ratio of reactants and products.

$Na_2CO_3 + 2HCl \rightarrow 2NaCl + H_2O + CO_2$
1 mole $Na_2CO_3 \equiv$ 2 moles HCl

Next calculate the number of moles of the sodium carbonate that were involved.

$n = C \times V$
$= 0.2 \times 0.025$ (remember volumes in dm^3)
= 0.005 moles Na_2CO_3

Then use the mole ratio from the balanced equation to work out the number of moles of the hydrochloric acid.

1 mole Na_2CO_3 = 2 moles HCl
0.005 moles $Na_2CO_3 = 0.005 \times 2 = 0.01$moles HCl

Finally calculate the concentration of the hydrochloric acid.

number of moles = concentration × volume
$n = C \times V$
$C = n/V$
$= 0.01/0.0074$
concentration of HCl = 1.35 mol dm^{-3}

D Analytical chemists, working for water authorities, use automatic titrations to carry out volumetric analysis.

3 A student used titrations to show that 25 cm^3 of sodium hydroxide was neutralised by 8.4 cm^3 of 0.2 mol dm^{-3} of sulphuric acid. What is the concentration of the sodium hydroxide?

$H_2SO_4 + 2NaOH \rightarrow Na_2SO_4 + 2H_2O$

4 Limewater is calcium hydroxide solution. The equation shows how it reacts with hydrochloric acid.

$2HCl + Ca(OH)_2 \rightarrow CaCl_2 + 2H_2O$

It was found that 100 cm^3 of the limewater needed exactly 20.0 cm^3 of 0.05 mol dm^{-3} hydrochloric acid for neutralisation.

a How many moles of calcium hydroxide are in 100 cm^3 of the solution?
b What mass of calcium hydroxide is in 100 cm^3 of the solution?
c What is the concentration of limewater in g dm^{-3}?

5 What information and titration measurements would you need to know to calculate the concentration of a solution of sodium hydroxide?

Summary Exercise | Higher Questions | Extension Questions

17. Questions

D

P

1 The testing of an unknown solution could involve both **quantitative** and **qualitative** analysis. Different practical techniques could be used including flame tests and **titrations**.
- **a** Explain the meaning of the three terms in **bold** in the above sentence.
- **b** What kind of substances are identified using flame tests?

2 Explain why at least two tests are needed to identify any ionic compound.

3 Name and describe the precipitate formed when:
- **a** sodium hydroxide solution is added to copper sulphate solution
- **b** silver nitrate solution is added to potassium bromide solution.

4 Write the formulae for the following ionic compounds:
- **a** calcium chloride
- **b** barium oxide
- **c** calcium sulphate
- **d** aluminium hydroxide.

5 When hydrochloric acid was added to a white powder the gas produced had a choking smell and turned blue litmus paper red. The powder produced a yellow colour when placed in a Bunsen flame. Suggest a name for this white powder.

6 A sample of water was tested to see what ions it contained. The results of the tests are shown below.

Test	Result
flame test	no colour produced
addition of sodium hydroxide solution	a white precipitate formed
addition of hydrochloric acid	no change occurred
addition of acidified silver nitrate solution	a yellow precipitate formed

What do these test results tell you about the water sample?

7 Calculate the formula mass of the following substances:
- **a** water (H_2O)
- **b** oxygen (O_2)
- **c** magnesium (Mg)
- **d** carbon dioxide (CO_2).

8 Calculate the number of moles in each of the following:
- **a** 16 g of helium gas (He)
- **b** 2.4 g of carbon solid (C)
- **c** 480 g of magnesium metal (Mg)
- **d** 10 g of calcium metal (Ca).

9 Calculate the volume of the following gases. (Molar gas volume = 24 dm^3.)
- **a** 1 g hydrogen gas (H_2)
- **b** 1.7 g ammonia (NH_3)
- **c** 70 g nitrogen gas (N_2)
- **d** 16 g ozone (O_3)

10 It was found that 200 cm^3 of a solution of sodium chloride contained 2.2 g of the salt. What was the concentration of this salt solution in g dm^{-3} and mol dm^{-3}?

11 Hydrochloric acid and magnesium react together according to the equation shown below:
$Mg + 2HCl \rightarrow MgCl_2 + H_2$
- **a** How many moles of hydrochloric acid are needed to react with 6 g of magnesium?
- **b** What mass of magnesium chloride will be formed from 6 g of magnesium?

12 Look at the equation below:
$Zn(s) + 2H^+(aq) + 2Cl^-(aq)$
$\rightarrow Zn^{2+}(aq) + 2Cl^-(aq) + H_2(g)$
- **a** Name the spectator ion(s) in this reaction.
- **b** Write the ionic equation for the reaction.

13 What volume of oxygen is needed to burn 40 cm^3 of propane?
$C_3H_8(g) + 5O_2(g) \rightarrow 3CO_2(g) + 4H_2O(l)$?

14 The purity of a water sample was tested by evaporating some of the water in a large evaporating basin and collecting the solid formed. It was found that when acid was added to the solid, effervescence occurred and the gas produced turned limewater milky. The solid also produced a dark red flame colour in a Bunsen burner flame.
- **a** What measurements would have to be taken to allow us to calculate the concentration of dissolved solids in g dm^{-3}?
- **b** Use the results to suggest the names of two ions that are present in the water sample.

15 What mass of sodium hydroxide (NaOH) is needed to make 250 cm^3 of 0.25 mol dm^{-3} solution?

16 The addition of sodium hydroxide solution to an 'unknown sample' can be used as a test for certain cations. For example if sodium hydroxide solution is added to copper(II) sulphate solution the following reaction occurs.

$CuSO_4(aq) + 2NaOH(aq)$
$\rightarrow Cu(OH)_2(s) + (Na)_2SO_4(aq)$

a What is this type of reaction called?
b What is a cation?
c What is the cation in copper(II) sulphate?
d How does the addition of sodium hydroxide identify the cation present?
e A similar reaction occurs between sodium hydroxide and iron(III) nitrate solution, $Fe(NO_3)_3(aq)$.
  i Write a balanced formula equation, including state symbols, to represent this reaction.
  ii What changes would you observe during this reaction?

17 The **volumetric analysis** of an acid usually involves a series of titrations using a **standard solution** of alkali. For example, standard sodium carbonate (Na_2CO_3) solution can be used to find the **concentration** of car battery acid, which is a solution of sulphuric acid (H_2SO_4).

$H_2SO_4 + Na_2CO_3 \rightarrow Na_2SO_4 + H_2O + CO_2$

The results of a titration of 10 cm^3 of car battery acid with 0.5 mol dm^{-3} sodium carbonate are shown below.

Titration	1st level (cm^3)	2nd level (cm^3)	Volume of Na_2CO_3 (cm^3)
1 (rough)	0.0	38.5	38.5
2	0.0	36.9	36.9
3	0.0	37.1	37.1

a Write definitions for all the words in **bold** in the above passage.
b What volume of sodium carbonate solution would you use in calculations to find the concentration of the acid?
c Use your answer to question **3b** to calculate the concentration of sulphuric acid in this car battery acid in mol dm^{-3}.
d Sulphuric acid reacts in a similar way with solid copper(II) carbonate ($CuCO_3$). In this reaction the salt copper(II) sulphate ($CuSO_4$) is produced.
  i What changes would you observe during this reaction?
  ii Write a balanced chemical equation, including state symbols, for the change.

18. Glossary

D

***acid** A solution which contains excess H^+ ions, and has a pH less than 7.
alkali A solution which contains excess OH^- ions, and has a pH greater than 7.
analytical chemist Chemist who investigates unknown substances to find out what substances are present and how much is present.
anion Negative ion formed when atoms or groups of atoms gain electrons.
***Avogadro's law** Rule which states that equal volumes of gases, at the same temperature and pressure, will contain equal numbers of molecules.
Avogadro constant 6×10^{23} mol^{-1}, the number of particles or 'formula units' in a mole of any substance.
balanced equation Formula equation having the same number of atoms of each element on both sides.
***base** Substance which neutralises acids, forming a salt and water.
burette Apparatus used to accurately measure the volume of solution that has been added during a titration.
cation Positive ion formed when atoms or groups of atoms lose electrons.
concentration The mass of solute in a given volume of solution, usually measured in g dm^{-3} or mol dm^{-3}.
covalent compound Compound formed between non-metals, by sharing pairs of electrons. It usually has a molecular structure.
diatomic molecule A molecule containing two atoms bonded together.
element A substance made up of only one kind of atom.
effervescence Reaction in solution which produces a gas.
***flame test** Test for metal ions using the colour they produce in a Bunsen burner flame.
forensic chemist Chemist who investigates the evidence from a crime scene to help prove guilt or innocence.
formula The ratio of the atoms of each element in a compound.
halide ion Ion of a group 7 element, for example fluoride (F^-), chloride (Cl^-), bromide (Br^-) or iodide (I^-).
***indicator** Substance which can change colour depending on the pH of a solution.
***ion** A particle, atom or group of atoms which has a positive or negative charge.
ionic compound A compound formed, usually between metals and non-metals, by the loss and gain of electrons.
***ionic substance** Substance which contains positive and negative ions, often formed between metals and non-metals.
ionic equation Equation which only shows the ions changed by a reaction, with the spectator ions omitted.
limewater Solution of calcium hydroxide in water, used to test for carbon dioxide, which turns it cloudy.
litmus A natural acid/alkali indicator, obtained from lichen. Turns red in acids and blue in alkalis.
mass A measure of the amount of matter a substance contains (units – grams (g) and kilograms (kg)).
***molar volume** Volume of a mole of any gas at a particular temperature and pressure.
***mole** The quantity of a substance which is equivalent to its relative formula mass in grams.
neutralise The reaction between an acid and a base that forms a salt and water.
pipette Apparatus used to accurately measure a set volume of a solution, which can be used in a titration.
pH scale Scale which commonly runs from 0 to 14 and measures the acidic or alkaline properties of a solution.
precipitate Solid formed in a reaction between two solutions.
***precipitation** A reaction between two solutions that forms a solid.
***purity** Measure of the amount of contamination in a sample of an element or compound.
***qualitative** Analysis which investigates the kind of substances present in a sample.
***quantitative** Analysis which investigates the amount of a substance present in a sample.
***reactant** The substances that take part in a chemical reaction.
relative atomic mass Average relative mass of the atoms of an element compared to a hydrogen atom.
relative formula mass Sum of the relative atomic masses for all the atoms of all the elements in a formula.
salt Compound containing ions which is formed during a neutralisation reaction.
solute Substance that dissolves in a solvent to form a solution.
solution Mixture of a solute dissolved in a solvent.
spectator ion Ion in a solution that remains unchanged during a chemical reaction.
standard solution Solution, of accurately known concentration, which is used in titrations, is stable when stored and produces a sharp colour change with the indicator.
***titration** Technique in volumetric analysis, used to find the exact volumes of solutions which react with each other.
universal indicator Mixture of dyes which forms a range of colours depending on the pH of a solution.
volumetric analysis Quantitative analysis which involves measuring the volumes of solutions.
volumetric flask Apparatus used to accurately make up an exact volume of a solution.

*glossary words from the specification

C3.4

Chemistry working for us

A Soap is an important product that is made using chemical reactions.

Almost everything in your life, from the food you eat to the clothes you are wearing and the paper and ink you will use in your lessons, will depend on the work of chemists.

Chemists try to make new substances or mixtures with improved properties. They may also find new ways of making substances more effectively or with less risk of pollution. While many chemicals are perfectly safe, some are not. Chemists also need to understand the dangers involved in working with chemicals, so that the risks to workers, consumers and the general public can be kept to a minimum.

In this topic you will learn that:

- chemistry is used in our everyday lives, for example in washing powders, sweets, cosmetics, paints, dyes and plastics
- the chemical and physical properties of elements and compounds are exploited to make useful and/or aesthetic products
- chemists are given a product specification and investigate which chemicals will be able to meet these requirements
- chemical substances need to be managed safely and considerately to ensure that they do not have a negative impact on the environment.

Look at these statements and sort them into the following categories:

I agree, I disagree, I can see both sides.

- Natural products e.g. wood and cotton are better than artificial ones e.g. plastics.
- All chemicals are compounds.
- Soap is better than shower gel for getting you clean.
- All batteries should be rechargeable.
- Chemists find most useful substances by accident.
- All chemicals are harmful.

V

D

1. Bells, pipes and wires

By the end of these two pages you should be able to:

- state the typical physical properties of transition metals such as iron and copper – high melting points, good conductors of heat and electricity and high density
- link these properties to the uses of these metals
- describe the uses of transition metals and their compounds as catalysts.

P

These tubular bells are made of steel. Many of their properties are typical of metals: they are shiny, strong, have a high melting point and they make a ringing sound when they are hit (**sonorous**). The bells are also very heavy to move around. This is because the metal has a high **density** – the mass of the metal is heavy for its size.

Metals can be bent into complicated shapes. The metal is said to be **malleable**. Examples include the tubes in a French horn, trombone or tuba.

A Tubular bells are shiny, sonorous and have a high density.

P

1 Name five properties of metals.
2 Look at the tubular bells. Which of the properties in question 1 are
   a essential
   b desirable?
3 Some people would say that metals are heavier than plastics. Why is 'denser' a better word to use than 'heavier'?

Pure iron is quite soft. It can be made harder and stronger by adding small amounts of other elements such as carbon to make steel. Steel is one of the most widely used metals in construction because of its strength and because iron is a common metal which is easy to obtain.

Metals are also **ductile** – they can be drawn out into thin wires. Electrical cables are often made of copper, because it is a good conductor of electricity. Metals are also good conductors of heat. Iron and copper are known as **transition metals**.

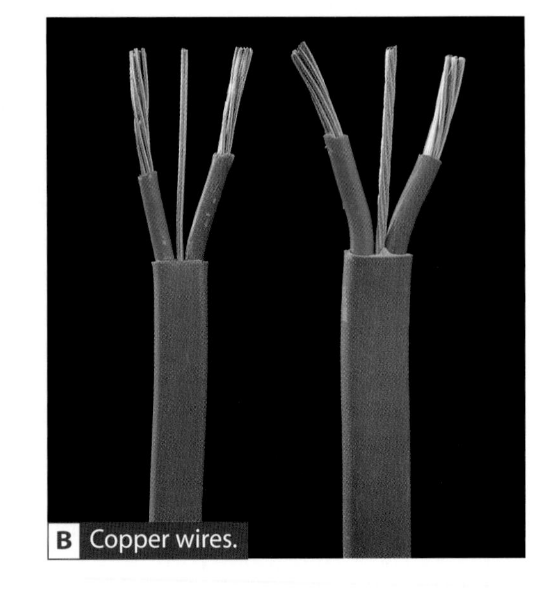
B Copper wires.

P

4 Name three transition metals other than iron and copper.
5 What **two** properties of copper make it suitable for electrical cables?
6 a Why are most electrical cables covered with an insulating layer of plastic (usually PVC)?
   b What other property of PVC makes it suitable for this purpose? Choose from: low density, low melting point, flexible, good heat insulator.

1	2											3	4	5	6	7	0
							H										He
Li	Be											B	C	N	O	F	Ne
Na	Mg											Al	Si	P	S	Cl	Ar
K	Ca	Sc	Ti	V	Cr	Mn	Fe	Co	Ni	Cu	Zn	Ga	Ge	As	Se	Br	Kr
Rb	Sr	Y	Zr	Nb	Mo	Tc	Ru	Rh	Pd	Ag	Cd	In	Sn	Sb	Te	I	Xe
Cs	Ba	La	Hf	Ta	W	Re	Os	Ir	Pt	Au	Hg	Tl	Pb	Bi	Po	At	Rn
Fr	Ra	Ac	Rf	Db	Sg	Bh	Hs	Mt	Ds	Rg							

C Transition metals are in the centre of the Periodic Table.

Transition metals are also used as **catalysts**. Platinum metal is used in the catalytic converter of a car exhaust. It speeds up the conversion reaction of the exhaust gases, but does not get used up in the reaction. Harmful gases such as carbon monoxide and nitrogen oxides are changed into nitrogen, water vapour and carbon dioxide which are much less harmful. A support material is coated with platinum, which is in the form of very tiny granules. Only a small amount of platinum is needed. The platinum can also withstand the high temperatures of the exhaust gases. Like most transition metals, platinum has a high melting point.

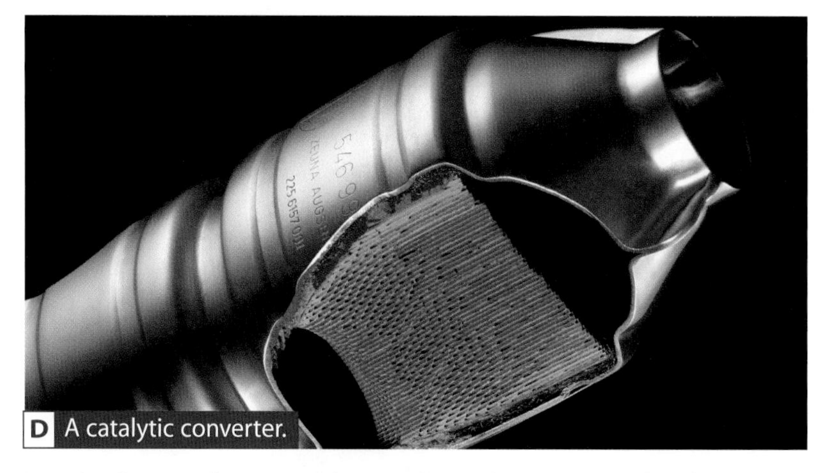

D A catalytic converter.

Iron is also used as a catalyst in the Haber process for the manufacture of ammonia. In this case the iron is in the form of a fine gauze.

7 Name two transition metals that may be used as catalysts.

8 Complete the following table with five examples drawn from this page or your general knowledge.

Name of transition metal	Use	Property of metal/ reason for use

Higher Questions

Extension Questions

2. Paints, dyes and pigments

By the end of these two pages you should be able to:

- describe the uses of transition metals and their compounds in pigments and dyes
- describe aspects of safety, sustainability and effects on the environment of pigments and dyes.

Have you ever wondered?

Which dye is used in denim?

You have seen that the transition metals are very useful to us as the pure elements, or as mixtures (alloys). The compounds of the transition metals can have important uses as pigments and dyes. This is because the compounds are often coloured. Compounds that are insoluble can be used as **pigments**, whereas soluble compounds are used as dyes.

Have you ever wondered?

How do paints get their colours?

The powdered pigment can be mixed with a liquid to form a suspension. This suspension can then be used as a paint. For example, white paint will often contain titanium dioxide as the pigment. Emulsion paints have the pigment dispersed in water together with a polymer such as PVC or acrylic resin. When the water evaporates, the pigment is left on the surface. Gloss paints use an oil-based liquid, which is why the brushes have to be washed with white spirit after use rather than water. The paint will include a solvent which will evaporate as the paint dries, and other chemicals that will harden as they react with the oxygen in the air (oxidise) to give the paint a hard-wearing finish.

1 How can you tell that the pigments in a can of emulsion paint are insoluble?

2 Name a transition metal which forms compounds which are:
  a white
  b blue
  c yellow.

3 Explain the difference between evaporation and oxidation. Which one is a chemical reaction?

A Artists use compounds of transition metals such as cobalt, titanium and chromium.

B Some soluble compounds of the transition metals.

Dyes are chemicals which are usually soluble and which will form a chemical bond to a fabric. Some dyes will bind directly to the fabric. Others require another chemical (called a **mordant**) to link the dye to the fabric.

Most of these dyes used to be made from natural organic sources such as petals, berries, crushed snails or even insects. It took a lot of these raw materials to make deep coloured dyes, and so it was only the very rich who could afford to wear colours such as purple and royal blue. These days, the wide variety of artificial dyes allows everyone to wear clothes that have bright colours.

The raw material for these dyes is likely to be coal or oil. Both of these are non-renewable resources which can also be used as fuels. The processes used to make the dyes involve some very corrosive chemicals and some of the compounds involved are **carcinogenic** – they can cause cancer. In the future we may have to depend more on renewable biomass (for example sugar, straw or wood) as a source of raw materials to make artificial dyes.

4 Look at photograph B. How can you tell that all the compounds in this picture are soluble?

5 What is a mordant?

6 Why is the deep blue colour known as *royal* blue?

7 Explain the difference between a dye and a pigment.

Summary Exercise

Higher Questions

3. Cosmetics and alcohols

By the end of these two pages you should be able to:

- describe the useful chemical and physical properties of alcohols, such as miscibility, as solvents
- describe the uses of alcohols in cosmetics.

Have you ever wondered:

What solvents are used in cosmetics?

A Alcohol is used as a solvent in many cosmetics.

Cosmetics contain a large number of chemicals. Some of these will be active ingredients, and others will be **solvents**, used to dissolve the different components. The commonest solvent is water, but water doesn't mix well with the oils that are often used in cosmetic products. In many of these products a type of **alcohol** is used as the solvent.

The word alcohol comes from the Arabic *al kohl* ('the spirit') and was used to mean a group of liquids that are 'spirits' which can be distilled from natural products. Now we use the word alcohol to mean carbon compounds which contain an OH (hydroxyl) group. A group of atoms like this is called a **functional group**. It can be found in many compounds and has characteristic properties. All alcohols have the functional group of an OH group bound to a carbon atom.

1. What is a solvent?
2. Some products list 'aqua' as the main solvent in the ingredients list.
   a What is 'aqua'?
   b Why is the common name not used on the bottle?
3. What solvent is often used to dissolve oily solids in cosmetics?

Alcohols are useful because they can mix with natural oils and with water. Oil and water do not mix together – they are **immiscible**. The alcohol allows us to make products where the components are **miscible** – they do not separate out into layers.

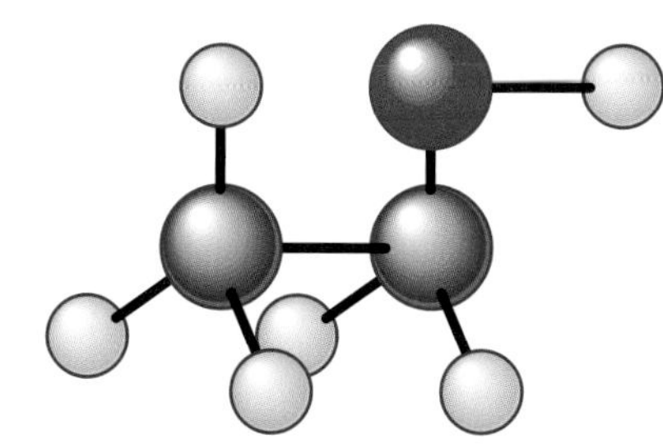

ethanol C_2H_5OH

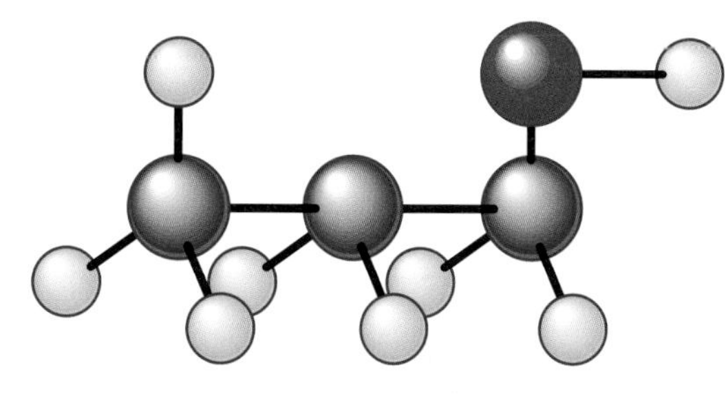

propanol C_3H_7OH

B The structure of ethanol (C_2H_5OH) and propanol (C_3H_7OH).

Two members of the alcohol family are ethanol and propanol. All alcohols have a name ending in '–ol'. The full name of the alcohol depends on the number of carbon atoms it contains. For example, hexanol contains six carbon atoms. The smaller molecules tend to mix very well with water. The longer the carbon chain, the less likely the alcohol is to mix well with water. Compounds which contain the same functional group but have a different carbon chain length are known as a **homologous series**.

4 Which three chemical elements are present in alcohols?

5 How are the structures of ethanol and propanol similar?

6 What is the difference between ethanol and propanol?

7 **a** What is the chemical formula of water?

b How is this similar to alcohols?

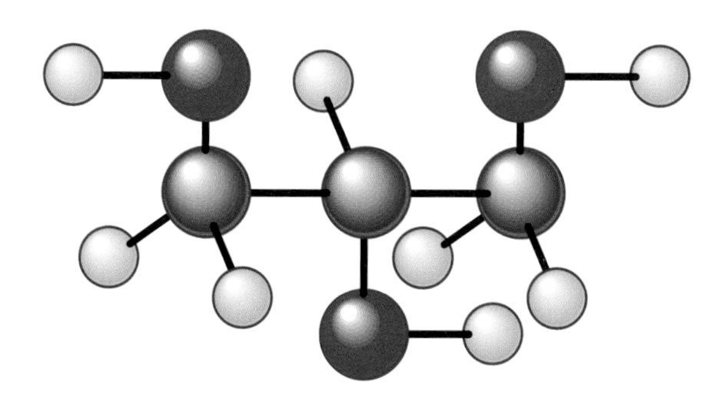

C Glycerol has three hydroxyl groups.

Glycerol is a particularly useful alcohol because it has three carbon atoms and three hydroxyl (OH) groups. This means that it mixes well with water and with a wide range of natural oils. Glycerol is a common ingredient in many creams and skin lotions.

8 Explain why we have to use a range of solvents in cosmetic products, rather than just relying on water.

Summary Exercise

Higher Questions

Extension Questions

4. Sweet and sour chemicals

By the end of these two pages you should be able to:

- describe the useful chemical and physical properties of organic acids, such as pH, acid behaviour and miscibility
- describe the useful chemical and physical properties of esters, such as miscibility, odour and use as a solvent
- describe the uses of alcohols in the preparation of esters
- describe the uses of esters in cosmetics and fruit flavourings.

If you leave a bottle of wine open to the air, it will 'go off'. The ethanol in the wine reacts with the oxygen in the air to make ethanoic acid. This is the active ingredient in vinegar. The word vinegar comes from the French *vin aigre*, meaning 'sour wine'.

A All these products contain vinegar (ethanoic acid).

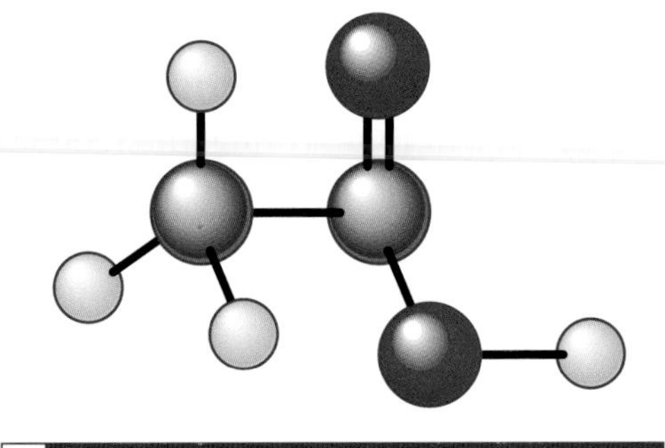

B The structure of ethanoic acid (CH_3COOH).

The ethanoic acid gives the food its tangy, sharp taste. The acid also acts as a preservative. The foods last longer because bacteria can't survive in the acidic environment.

Ethanoic acid is one of a group of **organic acids** which are also known as carboxylic acids. Organic acids are carbon compounds containing a carboxyl group (–COOH) as their functional group. Acids which do not contain carbon, such as hydrochloric acid and sulphuric acid, are called inorganic acids.

Pure ethanoic acid is very corrosive, but the vinegar that you put on your food is safe to taste because it is very dilute. Ethanoic acid is a weak acid – the **pH** is higher (typically 3 or 4) than many inorganic acids. It shows the normal properties of acids. For example, it will neutralise **alkalis** to form a salt and will react with magnesium to give off hydrogen.

1 Give two reasons why ethanoic acid is used in foods such as pickled onions.

2 What is the difference between an organic acid and an inorganic acid?

3 You put a piece of magnesium into a tube of ethanoic acid.
   a State one observation you would make.
   b Describe how the reaction might be different from that with a strong acid such as hydrochloric acid.

4 State one way in which the ethanoic acid molecule is similar to ethanol and one way in which it is different.

Have you ever wondered?

Which chemical substance smells like pear drops?

Organic acids will react with alcohols to form another class of substance called an **ester**.

For example:

ethanol (alcohol) + ethanoic acid (acid) → ethyl ethanoate (ester) + water

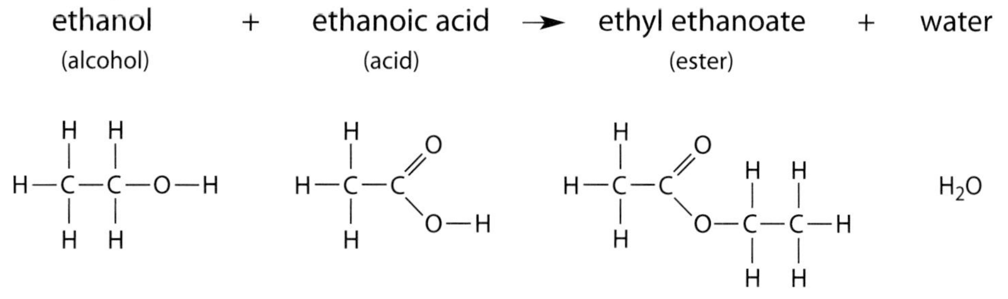

Esters are useful products because of their smell. Esters occur naturally in fruits, and we can use esters as **flavourings** in sweets such as pear drops or as scents in perfumes. Esters can also be used as solvents, for example in some nail varnish removers. Board markers contain an ester as a solvent. When you write on the board the solvent evaporates very quickly leaving the pigment behind. This can then be wiped off with a dry cloth when you have finished.

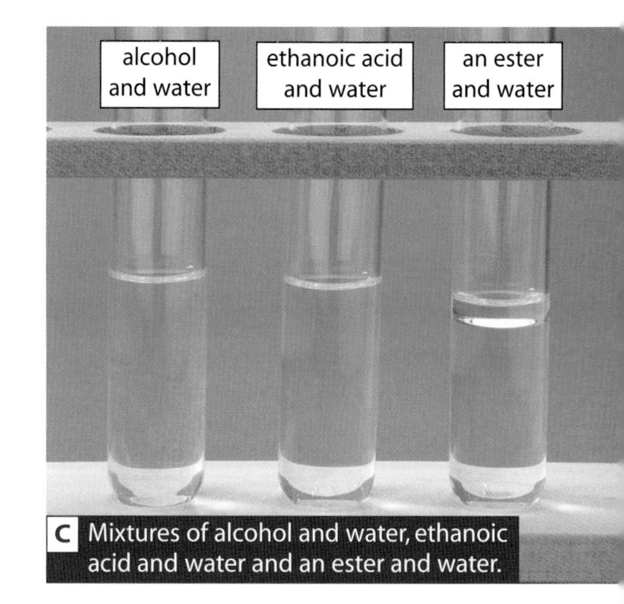

C Mixtures of alcohol and water, ethanoic acid and water and an ester and water.

Ethanol and ethanoic acid both dissolve well in water. These liquids are miscible. You can see in the photo that the liquids mix to give a single, clear solution. Esters do not mix with water – the liquids separate out into two layers. They are immiscible.

5 **a** Why does it tell you to put the top back on the dry wipe board marker as soon as you have finished with it?
b Why is this less of a problem with markers that use water as the solvent?

6 What does the word miscible mean?

7 Give one example of a pair of liquids which are:
a miscible
b immiscible.

8 Copy and complete the table to summarise the important properties and uses of alcohols, organic acids and esters.

Type of compound	Properties	Uses	Example
alcohol			
organic acid			
ester			

Summary Exercise

Higher Questions

Extension Questions

5. Cells and batteries

By the end of these two pages you should be able to:

- describe the types of cells used in electrical devices
- describe aspects of safety, sustainability and environmental impact in relation to the use of cells and batteries.

Have you ever wondered?

How do batteries work and what are the '+' and '–' poles?

It's difficult to imagine a world without battery-powered objects. We have come to depend on reliable sources of portable electrical power. Chemists have invented a range of batteries that are suitable for particular uses. They all have one thing in common – they transform the energy from chemical reactions into electrical energy.

Photograph A shows some different **cells**. The chemicals inside the cell will react so that it acts like a pump, pushing electrons around the circuit. A single cell will produce an electrical force of about 1.5 volts. Many devices will operate on a higher voltage than this, so you have to use several cells connected in series. A group of cells connected like this is called a **battery**.

Photograph B shows a battery. It contains a number of cells inside it. We often use the word battery when we actually mean a cell.

A A variety of cells.

B A battery.

1 State the energy transfer that takes place inside a cell.

2 What is the difference between a battery and a cell?

3 A torch needs two 1.5 V cells to make it light up. What is the voltage of the torch?

4 How many 1.5 V cells would there be inside a battery rated at:
a 6 V
b 9 V?

The first battery was invented in 1799 by Alessandro Volta. This was a pile of zinc and copper plates, separated by cloths that were soaked in salt water. A simple (wet) cell can be made by placing a zinc plate and a copper plate into a beaker of a conducting solution such as dilute sulphuric acid. A wet cell like this is not very practical as it is quite large and difficult to move without spilling the chemicals. The answer was to develop the 'dry' cell, which uses a moist paste instead of the conducting liquid.

5 Was Volta's pile a cell or a battery? Explain your answer.

6 Why is the 'dry' cell not really dry?

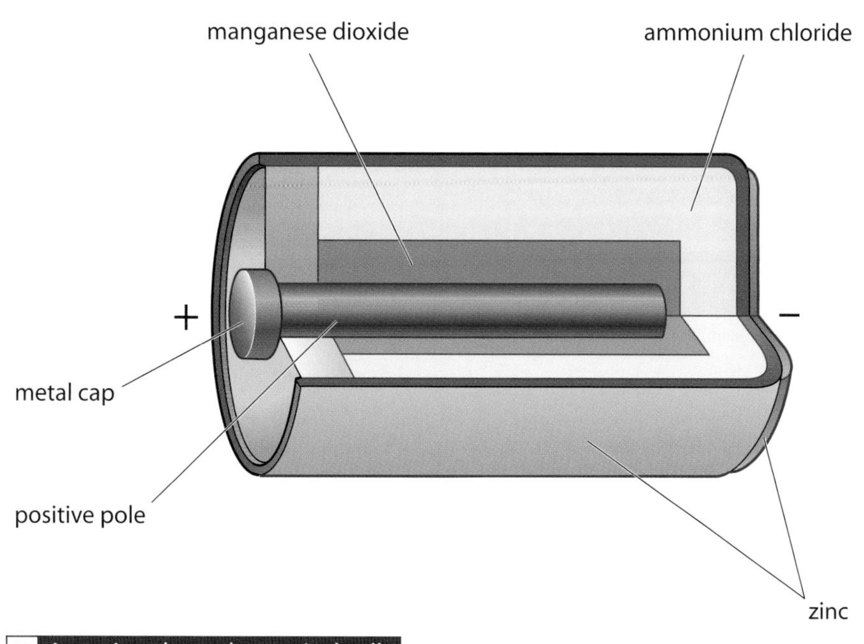

C A section through a typical cell.

Diagram C shows a typical dry cell. The case and bottom of the cell are made of a metal such as zinc. The zinc atoms will turn into ions and release some of their electrons which travel round the circuit. This is the negative pole, because electrons have a negative charge. After the electrons have travelled around the circuit through the wires, they enter the cell at the metal cap, then move down the positive pole of the cell. Around the positive pole there is a chemical such as manganese oxide that can absorb the electrons. Between the positive and negative poles, there is a paste of another chemical (the **electrolyte**) such as ammonium chloride. This is made of ions and is kept damp so that the ions can move through the cell. If the metal at the negative pole reacts too much the case can get too thin and the cell will start leaking. Alkaline cells use potassium hydroxide as the electrolyte.

Once all the chemicals have reacted, the cell is 'dead' and has to be thrown away. These single-use cells are called primary cells. Other cells are rechargeable (or secondary cells). This means that the chemical reaction can be reversed when the cell is connected to the mains supply. Although **rechargeable** cells are more expensive to buy, they work out cheaper in the long run because they can be used many times. Using rechargeable cells also reduces the amount of waste that is sent to landfill sites. Some cells contain toxic metals such as mercury and cadmium which can leak when the cells are thrown away.

7 **a** What problems could there be if a cell starts leaking?
b Why can't you solve this problem by making the cell completely dry?

8 A promotional offer says: 'Save money and save the environment. Buy this pack of two 1.5 V rechargeable batteries and save energy'.
a How does using rechargeable cells help the environment?
b Will you save money by using rechargeable cells? Explain your answer.
c Why is it incorrect to say that the pack contains 1.5 V batteries?

P

Summary Exercise | Higher Questions | Extension Questions

6. Electrolysis

By the end of these two pages you should be able to:

- explain the process of electrolysis
- describe types of electrolytes and electrodes
- describe the movement of ions during electrolysis and show simple circuits.

Have you ever wondered?

What is an electrolyte?

Electrical cells and batteries release the energy from chemical reactions in the form of electrical current. It is also possible to carry out this process in reverse. Electrical energy can be used to break the bonds between chemicals and split them up into elements. This process is called **electrolysis**.

A Salt water can be split up using electricity but solid salt cannot.

Photograph A shows the electrolysis of salt water. The current travels through the salt water and at the same time splits it up. A substance which is split up in this way is called an electrolyte. The chemical reactions occur at the two rods where the current enters and leaves the electrolyte. These rods are called **electrodes**. Metal electrodes will conduct electricity but may also take part in the reactions. Carbon electrodes do not usually react – they are **inert**.

During the electrolysis of salt water, hydrogen gas is formed at the negative electrode and chlorine is formed at the positive electrode. The electrolyte has to be in liquid form for the electrolysis to take place – solid salt is a non-conductor. Usually the electrolyte is dissolved in water to form an aqueous solution, but electrolysis will also take place if you heat the solid electrolyte until it melts.

1 Look at photograph A. What observation shows that:
  a salt water is a conductor of electricity
  b a chemical reaction is taking place?

2 When salt water is electrolysed, two products are formed at the electrodes.
  a Name the two products.
  b Which product comes from the salt and which from the water?

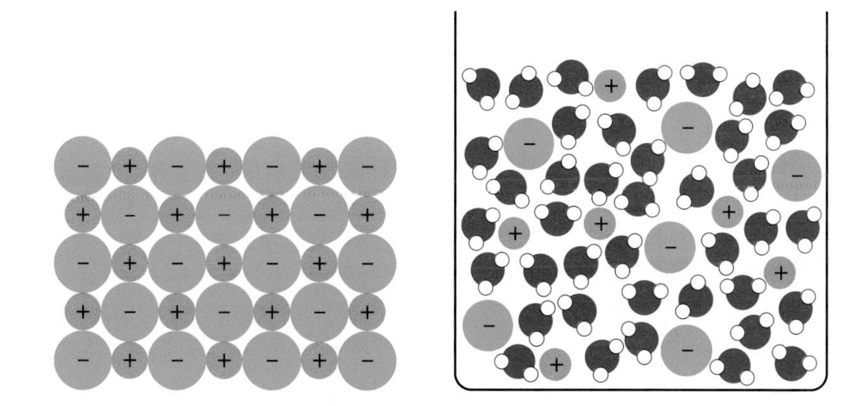

B Sodium chloride: as a solid and in solution.

Solid salt is made of a regular pattern of positive sodium ions and negative chloride ions. An **ion** is an atom which has a positive or negative charge. The oppositely charged ions attract each other strongly, forming a chemical bond. The energy from a battery is not enough to break the bonds in solid sodium chloride. The ions cannot move and the salt is a non-conductor. When salt dissolves in water, the water molecules surround the sodium and chloride ions, which can move around more easily. When a voltage is applied, the ions can move through the liquid and the solution conducts.

Metal ions are always positively charged. Non-metals are usually negative, except for hydrogen, which forms positive ions. This means that during electrolysis metals and hydrogen will be formed at the negative electrode (the **cathode**) and non-metals will be formed at the positive electrode (the **anode**).

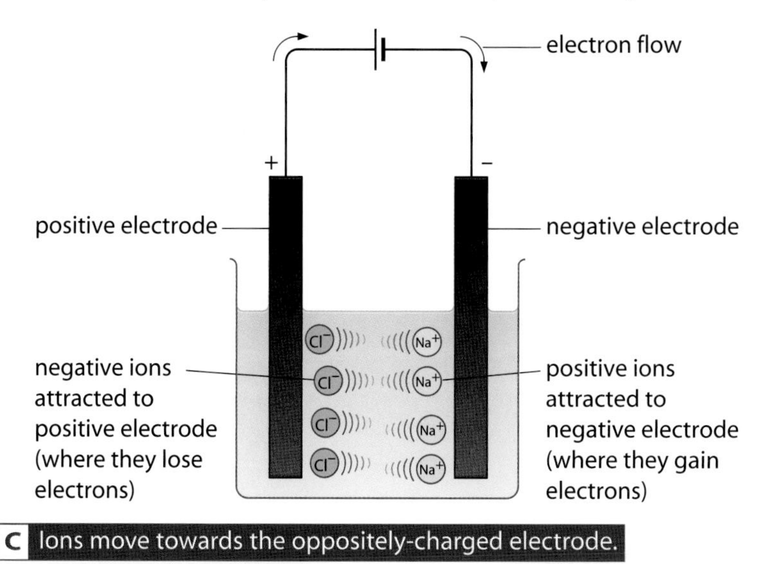

C Ions move towards the oppositely-charged electrode.

3 Do these symbols represent atoms or ions?

a Na
b Na^+
c H
d Br^-
e O^{2-}
f Ag^+

4 Why do metal ions move to the cathode during electrolysis?

5 When salt water is electrolysed, is chlorine produced at the anode or the cathode? Explain your answer.

6 Draw and annotate a diagram to show how aqueous copper chloride can be electrolysed to produce copper and chlorine.

P

A

Summary Exercise | Higher Questions | Extension Questions

7. Purifying and plating metals

By the end of these two pages you should be able to:

- describe the purification of copper by electrolysis
- draw a simple diagram to show the method used to purify copper.

A Copper wires do not contain many impurities.

Copper is a good conductor of electricity, but small amounts of impurities can reduce the conductivity significantly. When copper is extracted from its ores, it is not pure enough to be used for electrical work. Electrolysis can be used to purify the copper. It can also be used to recycle and purify scrap copper.

B Electrolysis on an industrial scale.

A block of impure copper is used as the anode (positive electrode) with copper sulphate as the electrolyte and pure copper as the cathode (negative electrode). The copper anode reacts by dissolving into the solution to form copper ions. An equal amount of copper is deposited from the electrolyte on to the cathode, which builds up a deposit of pure copper. Any impurities from the anode fall to the bottom of the liquid. This is not just useless waste, as it may contain some valuable metals such as silver and gold.

1 In the purification of copper by electrolysis, which substance is:
   a the anode
   b the cathode
   c the electrolyte?
2 What happens to any impurities?
3 Why is this waste not thrown away?

C These objects have all been electroplated.

Electrolysis is also used in electroplating, where a thin layer of one metal is coated on to another. The object to be plated is made the cathode in the electrolysis cell. The electrolyte is a solution which contains the metal to be plated. The anode will also be made of the metal which is to make up the coating.

Most cans are made of steel. The steel needs to be coated so that the iron does not react with the juices in the can and spoil the food. Tin is a less reactive metal than iron. If the steel is coated with tin there will be less of a reaction with the contents of the can. Copper coins are now mostly steel as well, with only a thin layer of copper on the surface. This means they are cheaper to produce. Recent 'bronze' coins will be attracted to a magnet, but older coins are not. All metals conduct electricity, but gold is a particularly good conductor. A thin layer of gold is plated on to steel for many electrical connections.

4 What is electroplating?
5 How you could use a magnet to tell the difference between a 2p coin from 1975 and a new one?
6 Why are headphone jacks not made out of pure gold?
7 Explain how a steel (iron) can could be electroplated with tin, using tin sulphate as the electrolyte. Include a diagram

Higher Questions

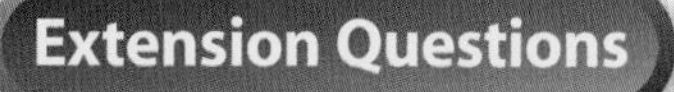

8. Electrolysis, ions and oxidation

By the end of these two pages you should be able to:

- understand what is meant by an ionic half-equation
- write ionic half-equations for simple electrolysis reactions
- describe oxidation as the removal of electrons
- describe reduction as the addition of electrons.

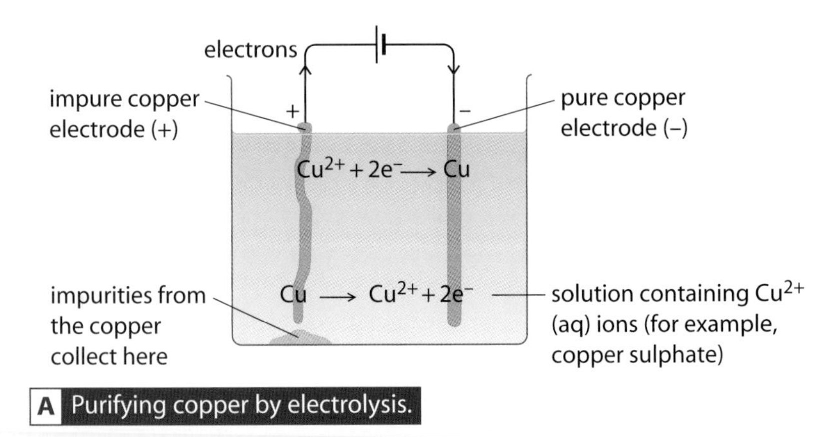

A Purifying copper by electrolysis.

Diagram A shows what happens when copper is purified using electrolysis. When the current flows, the positive metal ions from the solution are attracted to the cathode. The copper ion picks up two electrons from the negatively charged cathode, which has spare electrons. The ion turns into an atom and the solid copper is deposited on the cathode. The pure piece of copper gets bigger. We can summarise this reaction using an ionic half equation:

$$Cu^{2+}(aq) + 2e^- \rightarrow Cu(s)$$

copper ion → copper atom

1 Make a neat copy of the diagram, and add the labels: battery, anode, cathode, electrolyte.

2 Write the chemical symbol (including the state symbols) for:
a copper ions in solution
b solid copper.

At the same time copper atoms from the anode will dissolve into the solution as ions. This keeps the concentration of copper in the solution at a constant level. The atoms lose electrons when they turn into ions. The half equation for this reaction is:

$$Cu(s) - 2e^- \rightarrow Cu^{2+}(aq)$$

copper atom → copper ion

The copper atom loses two electrons, so the ion has a 2+ charge. The electrons from this reaction are pumped around wires in the circuit by the battery, from the anode to the cathode.

3 During the purification of copper by electrolysis what happens to:
a the mass of the anode
b the mass of the cathode
c the concentration of copper ions in solution?

4 What is the name of the particle which carries the electric current through:
a the wires in the circuit
b the electrolyte?

The burning of magnesium is an example of an **oxidation** reaction. The magnesium reacts with oxygen in the air to form magnesium oxide. When it does this, the magnesium atoms turn into magnesium ions by losing two electrons like this:

$Mg - 2e^- \rightarrow Mg^{2+}$

The electrons lost by the magnesium are picked up by the oxygen which turns into negative ions. Magnesium will also burn in other elements, such as chlorine, in a very similar way. We now use the word oxidation to describe *any* reaction in which electrons are lost, whether oxygen is involved or not.

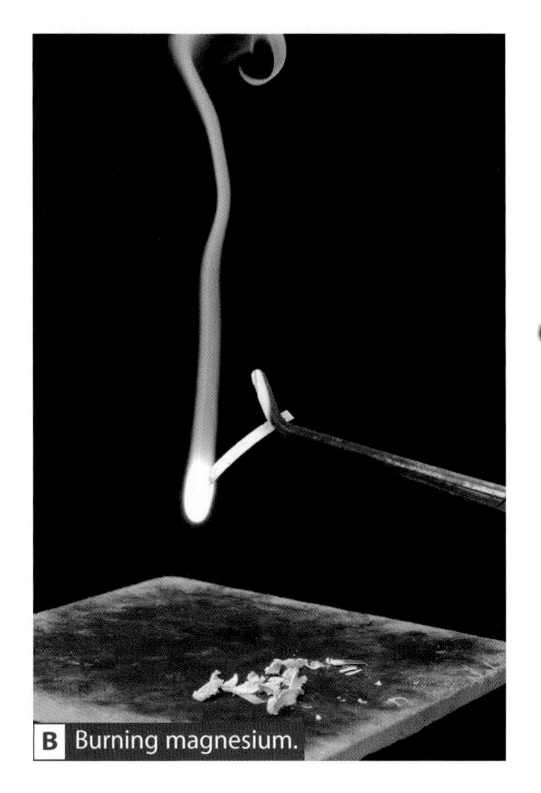

B Burning magnesium.

Both of these piles contain the same amount of iron. When a metal is produced from one of its compounds the mass of the metal is always less than the mass of the compound that you started with – the solid has reduced in mass. This is why the removal of oxygen or other non-metals is known as **reduction**.

C Iron oxide → iron: a reduction reaction.

When this happens, a metal ion is converted into a metal atom. In this case the half equation is:

$Fe^{3+} + 3e^- \rightarrow Fe$

Any reaction in which electrons are gained can be classed as a reduction reaction. The simple way to remember this is using the acronym OIL RIG [**O**xidation **I**s **L**oss (of electrons), **R**eduction **I**s **G**ain (of electrons)]. During electrolysis, oxidation occurs at the anode and reduction occurs at the cathode.

5 Write word equations for the reactions between:
a magnesium and oxygen
b magnesium and chlorine.

6 Classify each of these reactions as oxidation, reduction or neither:
a $Fe^{2+} + 2e^- \rightarrow Fe$
b $H^+ + OH^- \rightarrow H_2O$
c $O + 2e^- \rightarrow O^{2-}$
d $Fe^{2+} - e^- \rightarrow Fe^{3+}$
e $Cu^{2+} + e^- \rightarrow Cu^+$

Summary Exercise

Higher Questions

Extension Questions

9. The alkali metals

By the end of these two pages you should be able to:

- recall that alkali metals are soft and have comparatively low melting and boiling points
- state the reactions of the alkali metals (lithium, sodium, potassium) with water to form hydroxides and hydrogen gas.

The first tables of the elements were produced in about 1800 by John Dalton. They showed some familiar metals like iron and copper, but the **alkali metals** – lithium, sodium, potassium – were not shown. At that time the white solid compounds soda and potash were thought to be elements because no-one had managed to split them up.

Humphry Davy decided to use electrolysis to see if he could split up potash, a well known alkali. In 1807, he succeeded in carrying out electrolysis experiments on molten potash and on solid potash that was just moist enough to conduct. Davy reported that he had made 'small globules with a high metallic lustre … some of which burnt with explosion and bright flame … and others merely tarnished'. He was the first person to have seen potassium metal. A week later he did a similar experiment to extract sodium. The alkali metals had been discovered. What Davy had done was to convert the potassium ions into potassium metal like this:

$$K^+(aq) + e^- \rightarrow K(s)$$

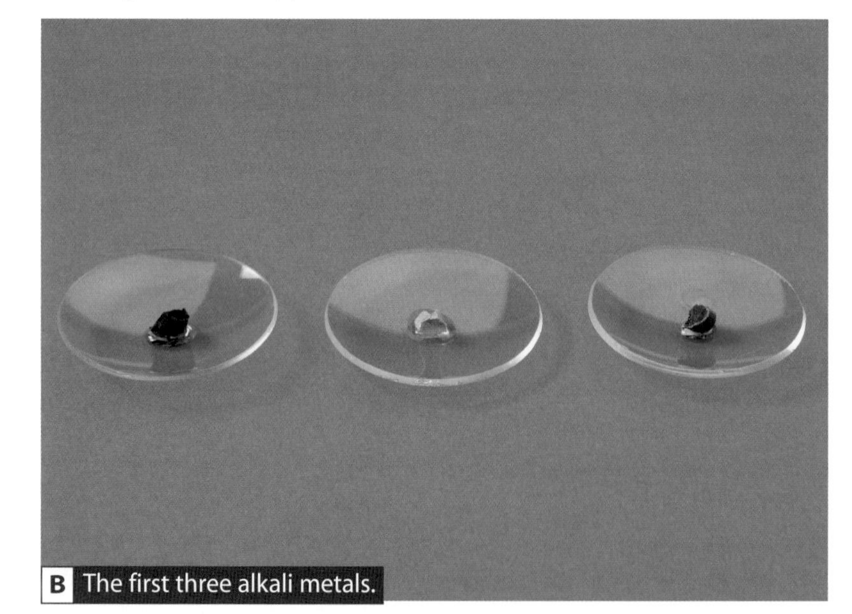

B The first three alkali metals.

The alkali metals do have some typical metal properties, but, compared to transition metals like iron and copper, the alkali metals are soft. All three can be cut with a knife. The melting points are also much lower than other metals.

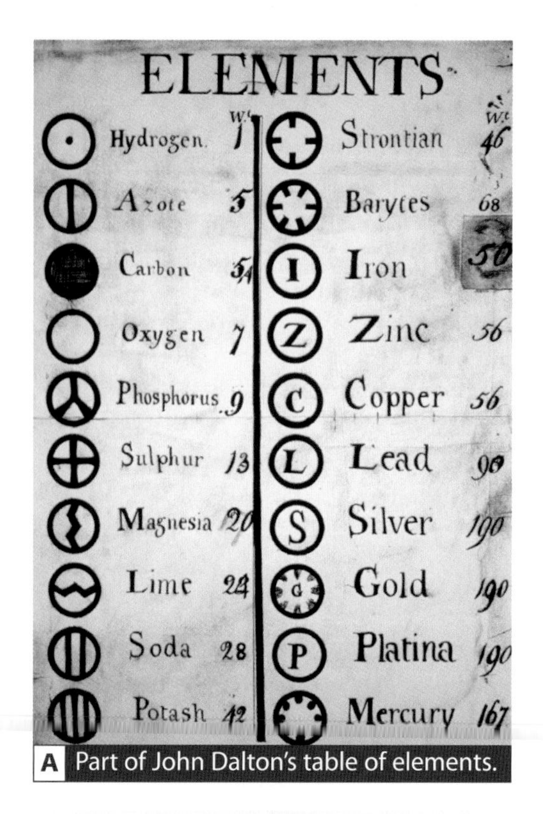

A Part of John Dalton's table of elements.

1 Caustic potash is potassium hydroxide.
  a Which three elements are present in potassium hydroxide?
  b Why was potash thought to be an element?
  c What observations made by Davy showed that he had made a new substance?

2 **a** Is the production of potassium metal from potash an oxidation or reduction reaction? Explain your answer.
  b The production of sodium is very similar. Write an ionic half equation for the change from sodium ions to sodium atoms.

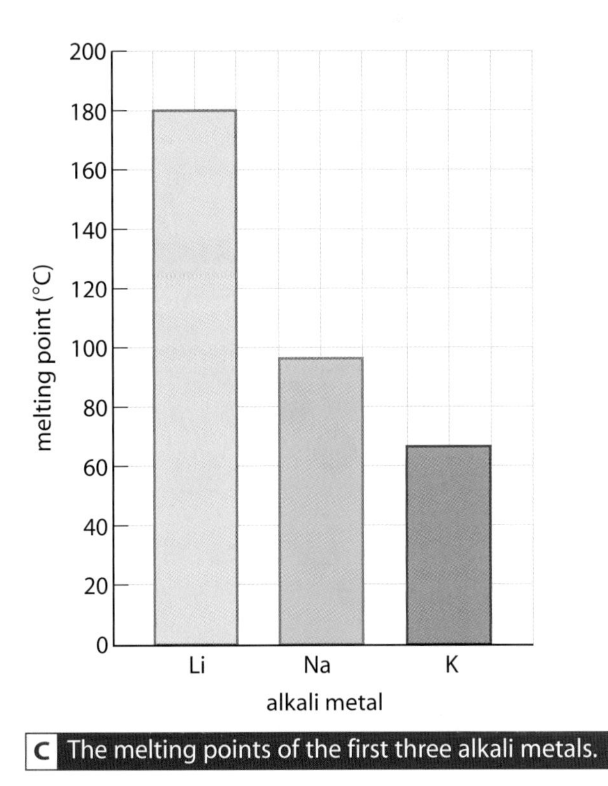

C The melting points of the first three alkali metals.

Alkali metals are found in Group 1 of the Periodic Table – they have one electron in their outer shell. The alkali metals are very reactive. When you put them in water, they react vigorously forming hydrogen gas and a metal hydroxide. For example:

sodium + water → sodium hydroxide + hydrogen

$2Na(s) + 2H_2O(l) \rightarrow 2NaOH(aq) + H_2(g)$

The alkali metals are too reactive to use safely: you can't use them to construct buildings or make cars. They can be useful in other forms, though:

- lithium ions are used in rechargeable batteries
- sodium vapour is used in street lights because it glows orange when a current is passed through it
- potassium ions are an important ingredient in fertilisers.

5 Write word and symbol equations for the reactions of lithium and potassium with water.

6 When the alkali metals react with water, what simple chemical test can be used to show:
a that hydrogen gas is produced
b that the solution produced is an alkali?

7 Explain why the alkali metals are stored under oil.

8 Summarise some of the similarities and differences between sodium (an alkali metal) and iron (a transition metal).

3 State one way in which the alkali metals look like typical metals.

4 **a** What is the melting point of lithium?
b Use the Periodic Table to place the atoms of lithium, sodium and potassium in order of size.
c What is the relationship between the melting point of the alkali metals and the size of the atom?
d The next member of the series is rubidium. Estimate the melting point of rubidium.

P

Summary Exercise

Higher Questions

Extension Questions

10. Glass

By the end of these two pages you should be able to:

- describe how sodium carbonate is used to make glass
- describe the uses of transition metals as pigments and dyes
- describe aspects of safety, sustainability and effects on the environment of glass.

Have you ever wondered?

How is glass made and coloured?

The alkali metals are very reactive. This means the pure elements have limited uses. However, there are many useful compounds of the alkali metals. One of these is sodium carbonate, Na_2CO_3 (soda ash, soda), which is used in the manufacture of glass.

The main raw material in glass manufacture is sand (silicon dioxide, SiO_2). To make the glass, you need to heat the sand until it melts. The problem is that silicon dioxide (also known as silica) has a giant structure with very strong bonds between the atoms.

Pure silica melts at about 2000 °C. Adding sodium carbonate ('soda') lowers the melting point to about 1000 °C. This soda glass is much easier to make, but it is not very strong and can even dissolve in water! Adding a little lime (calcium oxide) improves the strength and durability of the glass. This soda-lime glass is the most common type of glass that we use.

B Some different types of glass.

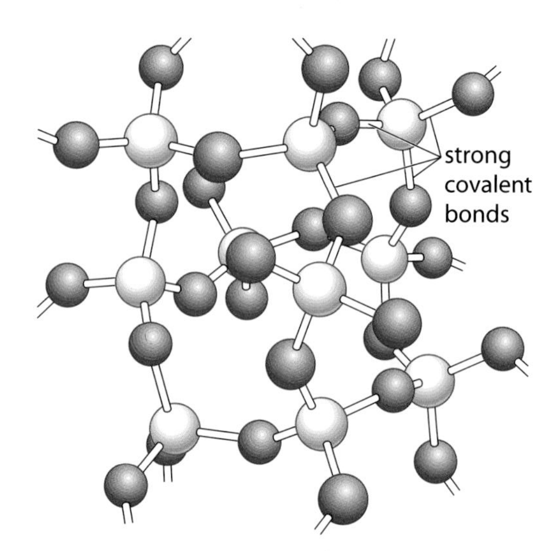

A The structure of sand crystals made from silicon dioxide (SiO_2).

1 What is the main raw material used to make glass?

2 Why are the following additives used in glass manufacture?
a soda
b lime

3 In the molecular diagram, which colours represent silicon and oxygen? How can you work this out?

The tumbler on the left in photo B is made of normal soda-lime glass. The crystal glass tumbler on the right contains different additives – potassium oxide and lead oxide. This changes the way in which the glass refracts light, giving it a sparkling appearance.

Normal glass cracks if you heat it, but Pyrex® glass is specially designed to be heat resistant. Boron is added to the glass, and this means it has a very low expansion rate, preventing the strains and cracks from forming.

Other specialist types of glass include:

- fibre glass, for loft insulation and lightweight structures
- glass ceramic, for cooker tops
- crown glass, for high quality lenses.

Transition metal compounds can be used to give colours to glass. Often very small amounts of these additives (typically between 0.01 and 2%) can give very rich colours.

Additive(s)	Colour of glass
iron(II) oxide and chromium	dark green
copper oxide	turquoise
cobalt	blue
copper or gold	deep red

Sand, the raw material used to make glass, is non-renewable, but there are very large supplies available, and so it is very unlikely that we will run out. Many glass objects such as bottles can be reused or recycled and glass factories can use recycled glass, as well as sand, as a raw material. Some countries now have a compulsory deposit on glass bottles to encourage re-use. You get your deposit back when you take the bottle back to the shop. The bottle is then transported back to the factory, cleaned thoroughly and refilled.

As plastic containers have become more common the use of glass has decreased. Neither glass nor plastic is biodegradable but glass can be re-used more easily. Concerns about the impact of plastics on the environment mean that glass may become more popular again in the future.

4 Name an element that might added to the glass used in:
a a green bottle
b an oven-proof dish
c a blue panel on a stained-glass window
d a crystal chandelier.

5 Explain the difference between recycling and re-using glass bottles.

6 What safety and environmental problems can arise if you just throw a glass bottle away?

7 Bottles can be cleaned and sterilised using steam. Why is this easier for glass bottles than plastics?

8 Bottle banks often have a section for 'clear' glass. What does 'clear' mean? Why is 'colourless' a better word than clear to describe this type of glass?

9 Copy and complete the table to show the elements that are used as additives in three different types of glass and three different *colours*.

Type or colour of glass	Elements added to the glass

Higher Questions

11. Caustic soda: an important alkali

By the end of these two pages you should be able to:

- describe the process of electrolysis
- describe the uses of sodium hydroxide to illustrate its economic importance in producing fibres.

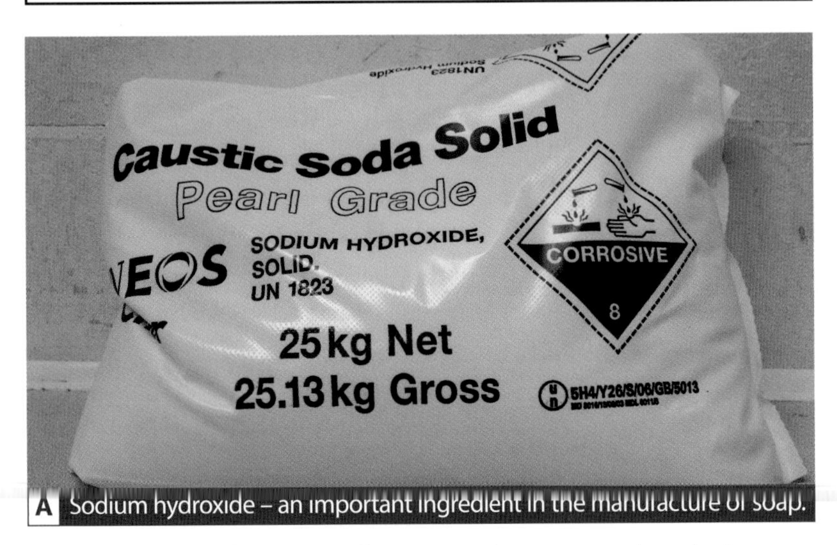

A Sodium hydroxide – an important ingredient in the manufacture of soap.

Sodium hydroxide is one of our most important chemicals. About 50 million tonnes is produced worldwide every year, with about 10 million tonnes being produced in Europe. Sodium hydroxide is a strong alkali which is also known as caustic soda. The word caustic means corrosive.

Sodium hydroxide is produced by the electrolysis of salt water (brine). Photograph B shows the reactions that take place. Universal indicator has been added to the salt water to show the reactions more clearly.

B Electrolysis of sodium chloride solution, with the anode and cathode compartments separated.

1 How much sodium hydroxide is produced in the world each year?

2 What percentage of this production takes place in Europe? Show your working.

3 Another important alkali is caustic potash. What is its proper chemical name?

Sodium hydroxide is formed at the cathode. The positive sodium ions are attracted to the negative electrode. Water molecules are decomposed in this reaction.

$$2Na^+(aq) + 2H_2O(l) + 2e^- \rightarrow 2NaOH(aq) + H_2(g)$$

At the anode, chlorine gas is produced. Some of the chlorine reacts with the water to make it acidic.

$$2Cl^-(aq) - 2e^- \rightarrow Cl_2(aq)$$

The chlorine can also act as a bleach, taking the colour out of the indicator altogether. The two halves of the reaction have to be kept separate, otherwise the chlorine and sodium hydroxide would react together. On an industrial scale this is done using a steel gauze or a porous membrane – in the laboratory we can use a piece of damp filter paper between the dishes. An industrial electrolysis plant will use very high currents – sometimes up to 100 000 amps. It takes about 3 units (kWh) of electrical energy to produce 1 kg of sodium hydroxide.

The sodium hydroxide formed has many important uses, including making soap, oven cleaners and in the manufacture of paper and viscose fibres.

C Sodium hydroxide can help to turn these logs into paper and fabrics.

Viscose (or rayon) is a clothing fibre that is made from the plant material cellulose. One important source of cellulose is wood. Wood is a natural product, but you can't make clothes out of wood directly. Sodium hydroxide can be used to react with the wood and break down some of the chemicals that make the wood tough. The cellulose can now be treated to turn it into a purer form which can be spun into fibres which are known as Rayon or viscose (*vis*cous cellu*lose*). Cellulose and viscose are both polymers – long-chain molecules with repeating units.

A similar process can be used to turn wood into pulp which can then be used to make paper.

4 In the electrolysis of brine, name the gas given off at:
a the anode
b the cathode.

5 What observation tells you that an alkali is being formed at the cathode?

6 Name three products made using sodium hydroxide

7 Give an example of:
a a natural fibre
b an artificial polymer used in clothing.

8 Explain why viscose doesn't fit exactly into either of these categories.

9 Explain why sodium hydroxide is the most widely produced alkali. Refer to the raw materials used and the uses for the product.

Summary Exercise

Higher Questions

Extension Questions

12. Why is sulphuric acid so important?

By the end of these two pages you should be able to:

- describe the uses of sulphuric acid to illustrate its economic importance in producing fertilisers and paints.

About 200 years ago, sulphuric acid (H_2SO_4) was considered to be one of the most important chemicals that a country produced. The same is true today – sulphuric acid ranks number one in the league table of manufactured chemicals, with over 170 million tonnes being produced every year across the world. It is important because it has such a wide range of uses, and many of these are of vital importance. Typical uses of sulphuric acid are shown in diagram B.

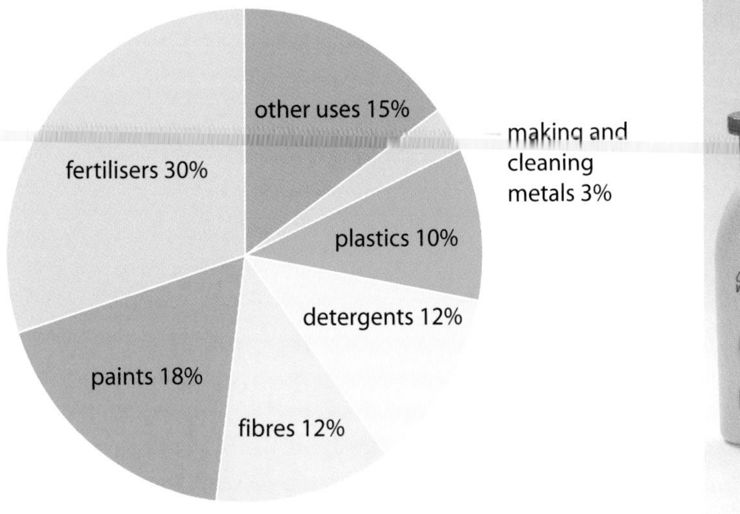

A Uses of sulphuric acid.

B These products are all made using sulphuric acid.

The most important elements in fertilisers are nitrogen, potassium and phosphorus, so the use of sulphuric acid may not be immediately obvious. It is used in two important ways:

- to make superphosphate fertiliser from rock phosphate (calcium phosphate)
- to make ammonium sulphate, a fertiliser which contains nitrogen.

1 List the four main uses of sulphuric acid.

2 Look back at the production figure for sodium hydroxide on the previous spread. How many tonnes of sulphuric acid are made for each tonne of sodium hydroxide?

3 Fertilisers often show their NPK ratio. Which elements do these three symbols represent?

Plants take in minerals through their roots. Calcium phosphate is insoluble, so it is no use as a fertiliser even though it contains phosphorus. When it reacts with sulphuric acid it is converted into calcium hydrogen phosphate like this:

$$Ca_3(PO_4)_2 + 2H_2SO_4 \rightarrow Ca(H_2PO_4)_2 + CaSO_4$$

The phosphorus is now in a soluble form which can be absorbed by the plant. This is known as superphosphate fertiliser.

Ammonia (NH_3) is a gas. Even though it is very soluble in water, if you sprayed ammonia solution on to a field a lot of the ammonia would be wasted by evaporation before it got into the plant. There would also be the problem of the smell. It is therefore much more convenient to convert it into ammonium sulphate, which is a soluble solid which is easier to handle with less waste.

4 Copy and complete the following word equations:

a calcium phosphate + __________ __________ → __________ hydrogen phosphate + calcium __________

b ammonia + sulphuric acid → __________ __________

5 Explain why rock phosphate and ammonia are not very good to use as fertilisers.

Sulphuric acid is also used in paint manufacture. Titanium dioxide is the pigment in white paint. The natural ore ilmenite contains titanium, but it also contains iron. The formula is $FeTiO_3$. To extract the titanium, the ilmenite is reacted with sulphuric acid to make iron sulphate ($FeSO_4$) and titanyl sulphate ($TiOSO_4$). The iron and any other impurities can then be separated from the titanium. Finally, the titanyl sulphate is heated to a high temperature with steam which allows the formation of pure titanium dioxide:

titanyl sulphate + steam → titanium dioxide + sulphuric acid

6 What is the formula of titanium dioxide?

7 What would be the problem with the pigment if there were impurities of iron in the titanium dioxide?

8 Summarise the uses of sulphuric acid by turning the information in diagram A into a table or bar chart, and annotating your chart with examples from this page or other sources.

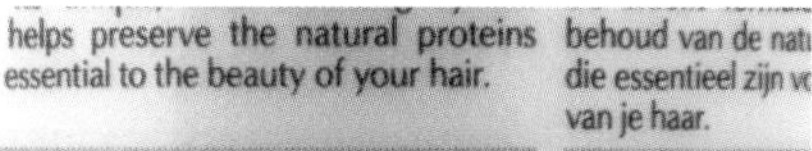

C Sulphuric acid is also used in the manufacture of detergents. This shampoo contains sodium laureth sulphate as one of the main ingredients. Sulphuric acid is also used in the manufacture of dyes, medicines and plastics. It is also the acid that is used in car batteries.

Summary Exercise | Higher Questions | Extension Questions

13. The Contact process

By the end of these two pages you should be able to:

- describe the manufacture of sulphuric acid from sulphur and sulphide ores
- describe the operating conditions used in the Contact process
- understand how sulphuric acid production is influenced by economic, safety and environmental factors.

Sulphuric acid is manufactured in the **Contact process** in a way which is as efficient and safe as possible. The product has to be sold at a price which is attractive to the customer, but of course the manufacturing company needs to make a profit as well.

The Contact process has four stages:

- production of sulphur dioxide
- conversion of sulphur dioxide to sulphur trioxide
- absorption of sulphur trioxide (to form oleum)
- dilution of oleum to form sulphuric acid.

A Manufacturing sulphuric acid.

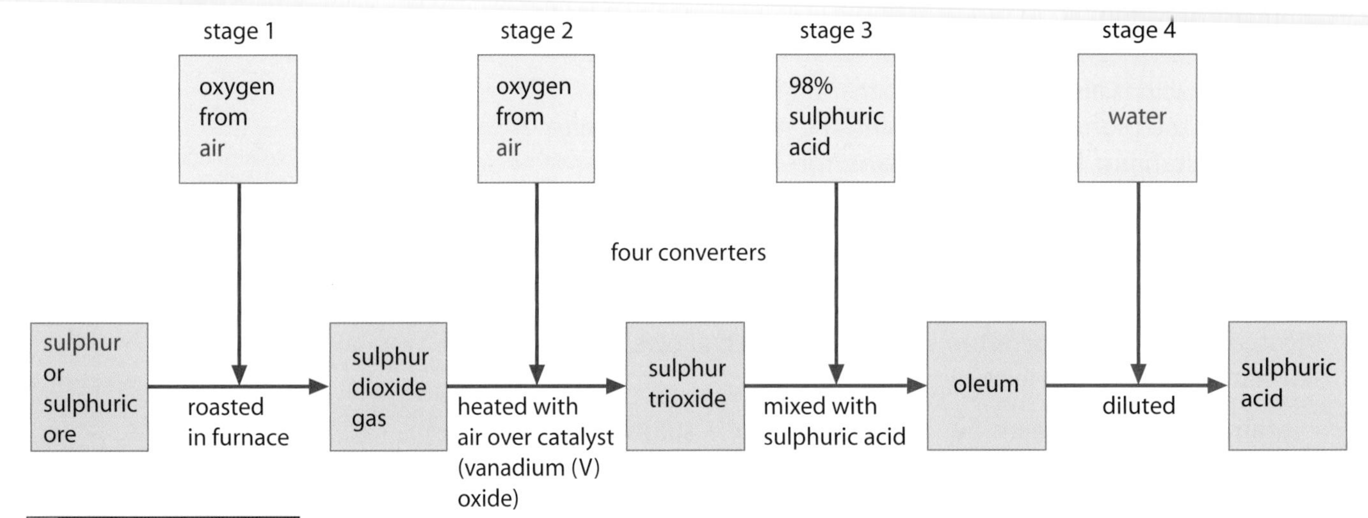

B The Contact process.

Sulphur dioxide can be made by burning sulphur in air.

$S(s) + O_2(g) \rightarrow SO_2(g)$

An alternative method involves roasting sulphide minerals such as iron sulphide.

iron sulphide + oxygen → iron oxide + sulphur dioxide

These reactions both release a lot of heat energy – they are strongly exothermic. The next stage of the process is the most difficult. Sulphur dioxide is converted into sulphur trioxide:

$2SO_2(g) + O_2(g) \rightarrow 2SO_3(g)$

1 What product is made in the Contact process?

2 Name the three raw materials that are used.

3 Suggest two reasons why the process needs to be as efficient as possible with the greatest possible yield of product.

The conversion of reactants to products is never 100% and the control of reaction conditions is crucial. A catalyst (vanadium oxide) is used to speed up the reaction. The gases are passed over layers of catalyst at about 450 °C. This is a compromise – the temperature is high enough to give a fast reaction, but not too high because very hot temperatures reduce the overall conversion rate. The gases are passed through three or four converters, one after another, to get the best overall conversion rate.

These reactions are also exothermic. The heat from the reaction is used to warm up the gases coming into the reactor. In this way the reactor can be kept at 450 °C without having to heat it up.

4 Why is a catalyst used in the reaction?

5 **a** What happens to the speed of a reaction if you increase the temperature?
b Use the graph to explain why the reaction is carried out at 450 °C rather than a higher temperature.

6 **a** Plot a graph of overall conversion rate against the converter number.
b Explain why one converter would not be enough, but there is no point adding more than four converters.

Converter number	Overall conversion (%)
1	63
2	84
3	93
4	99.5

C Shower gel, shampoo and foam bath are all types of detergent.

The reactor is operated slightly above normal atmospheric pressure. Higher pressures would give an even better conversion, but the cost, safety problems and increased corrosion of the pipes that would occur make this uneconomic. Finally, excess air is used to reduce the amount of unreacted sulphur dioxide, as this would cause a pollution problem. 99.5% conversion is now the minimum efficiency allowed by law.

The last stage of the reaction involves dissolving the sulphur dioxide in water.

$SO_3(s) + H_2O(l) \rightarrow H_2SO_4(l)$

This reaction is so exothermic that the heat can cause the sulphuric acid to boil. For safety reasons, the sulphur trioxide is reacted first with 98% sulphuric acid to form a liquid called oleum. This can be diluted in a controlled way to make more sulphuric acid. The acid is then stored in tanks.

7 What safety problems might be caused by:
a operating the plant at too low a conversion rate
b using too high a pressure
c dissolving the sulphur trioxide directly in water?

8 Make a summary of the reaction conditions used in stage 2 using the sub-headings: temperature, pressure, catalyst, ratio of air to SO_2.

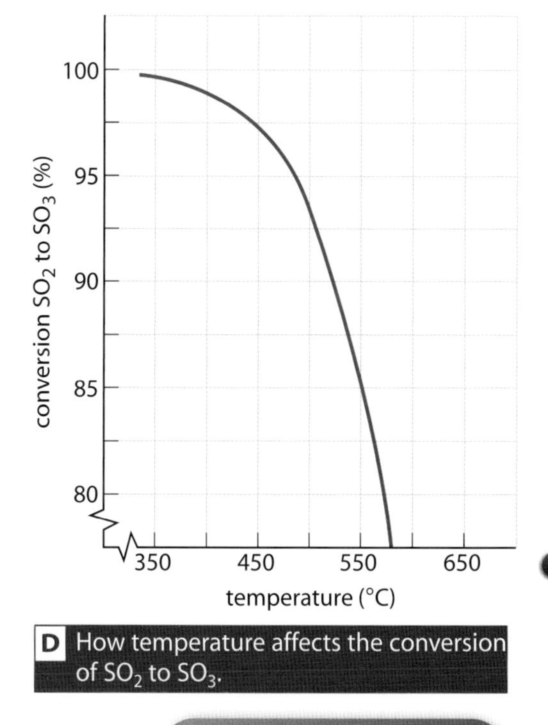

D How temperature affects the conversion of SO_2 to SO_3.

Summary Exercise

Higher Questions

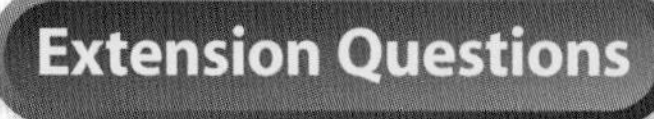

14. Detergents

By the end of these two pages you should be able to:

- describe the uses of sulphuric acid to illustrate its economic importance in producing detergents
- explain the detergent action of surfactants in lowering surface tension to remove dirt and/or oil/grease
- describe the differences between 'biological' and 'non-biological' detergents
- consider the effects of detergents on the environment.

A Shower gel, shampoo and foam bath are all types of detergent.

Have you ever wondered?

How do detergents remove fats or dirt from clothes?

A **detergent** is a substance that cleans. In one sense water is a detergent as it is a good solvent. However, water does not always wet things very easily. Sometimes it will form droplets that sit on the surface of a fabric and do not mix in. Water will not mix at all with oils, so on its own it is not very good at removing greasy dirt. Detergents can range from washing-up liquid and kitchen surface cleaners to shampoo, soap and toothpaste.

1 Explain why water on its own is not very good at cleaning off dirt.

2 What would you observe if you put some water into a tube with some cooking oil, shook it up and left it to stand?

Diagram B shows a detergent that is made using sulphuric acid. The carbon chain usually comes from petroleum products, although some detergents use plant oils as a source of carbon.

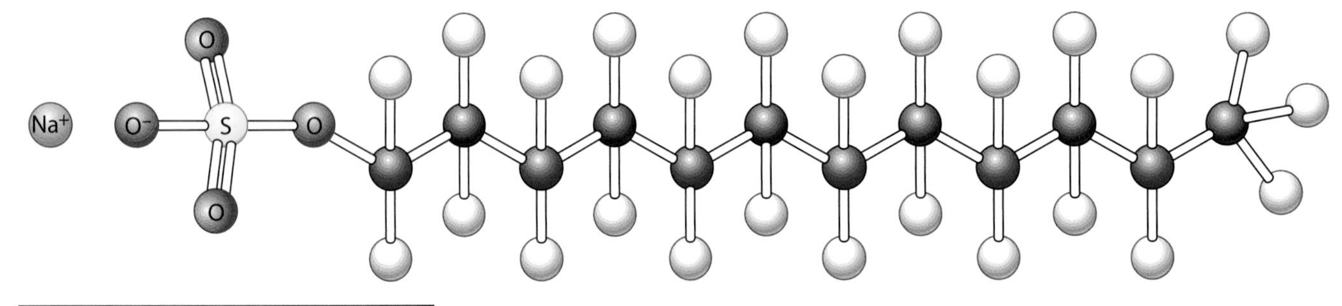

B Molecular diagram of a detergent.

Water molecules have a force of attraction for each other. This is not as strong as in a solid like salt, but it is still strong enough to hold the water together. On the surface of the water droplet the overall force acts inwards, pulling the water into a drop shape. This force is known as **surface tension** – it is like a skin on the surface of the water. A detergent is a chemical which can reduce the surface tension of the water and break up the droplets. It is an example of a **surfactant** ('surface active agent').

Have you ever wondered?

What is meant by 'hydrophilic' and 'hydrophobic'?

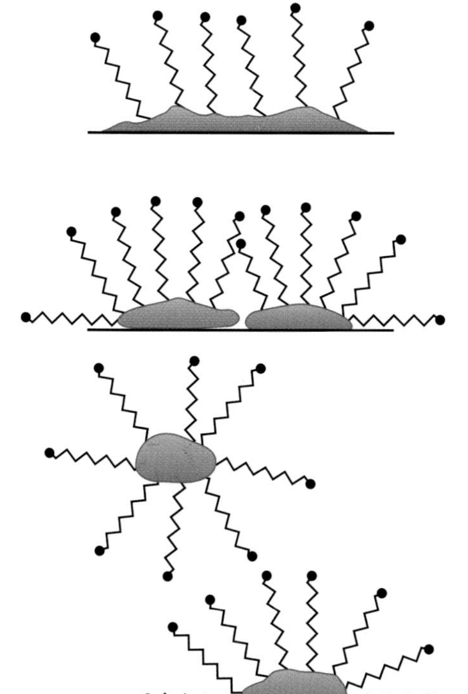

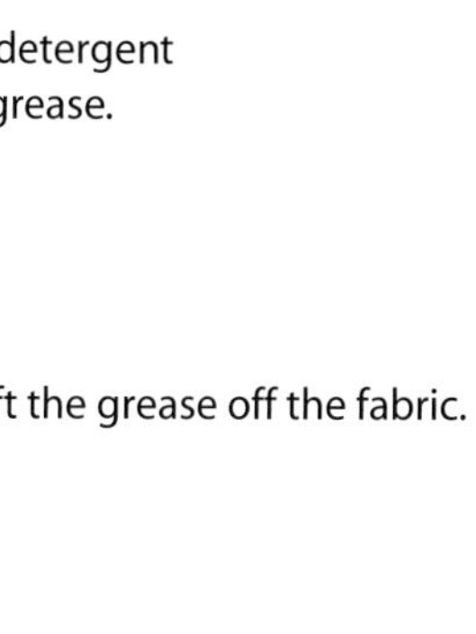

C How a detergent works.

A model of a detergent molecule often shows it as a 'tadpole'. The 'head' usually contains oxygen atoms, which mix well with water. This is the **hydrophilic** (water-loving) part of the molecule. This end may also be charged, in which case the detergent is **ionic**. The 'tail' is a long chain of carbon atoms which mixes well with oils. This end does not mix well with water – it is said to be **hydrophobic** (water-hating). The detergent molecules can get in between the water molecules and reduce the forces of attraction between them – this means that the surface tension is reduced.

When the detergent is washed down the drain, the carbon chains are broken down by bacteria in the water. Some of the early synthetic detergents had molecules which were non-biodegradable, which caused big pollution problems on rivers, with foam forming on the surface of the water. Now, all detergents on sale in Britain have to be biodegradable.

Some detergents have enzymes added to them. These help the detergent to break down stains that have a high protein content, such as egg and blood. They are called biological detergents because they contain enzymes ('biological catalysts'). Some people find that the enzymes in the detergent will react with their skin causing an allergic reaction, so they prefer non-biological products.

A

3 Draw a diagram of a detergent molecule and label the hydrophilic and hydrophobic sections.

4 What does 'biodegradable' mean? Why is it important that detergents that we use in the home are biodegradable?

5 What are biological detergents? Why are they called this?

P

6 Imagine you were trying to explain to a much younger child how detergents work. Invent a detergent 'character' and draw a cartoon strip or make up a simple picture story to explain the ideas in a way that a six or seven year old might understand.

Summary Exercise | **Higher Questions** | **Extension Questions**

15. Soap

By the end of these two pages you should be able to:

- describe how soap is made from fatty acids and an alkali
- describe the uses of organic acids in soaps and detergents
- describe the uses of sodium hydroxide to illustrate its economic importance in producing soaps.

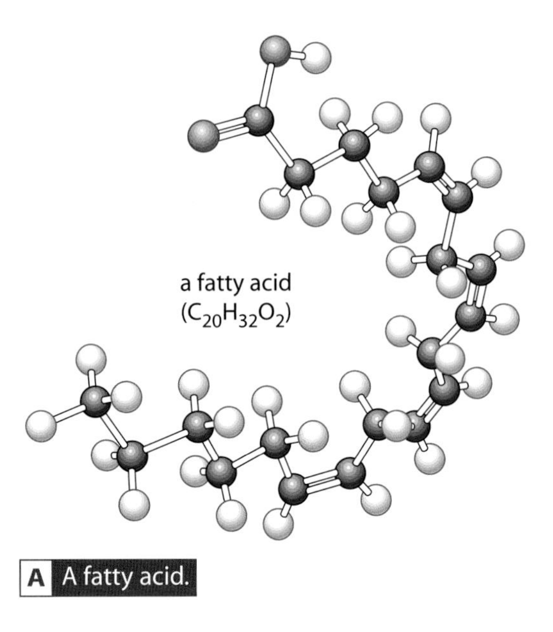

A A fatty acid.

Soap is a detergent that has been known for over 3000 years. The raw materials are fats or oils and an alkali such as sodium hydroxide. In ancient times the fat would have probably come from an animal such as a sheep. The alkali would have been made from the metal oxides found in the ash at the end of a fire. The word alkali is derived from the Arabic *al kali*, meaning 'the ashes'.

1 What are the two types of ingredient needed to make soap?

Fats and oils are esters. They are made from alcohols and **fatty acids** – organic acids which contain a long chain of carbon and hydrogen atoms and a carboxylic acid group (COOH). Diagram A shows a typical fatty acid.

2 What type of compound are fats and oils?

3 Name the three chemical elements present in a fatty acid.

4 Copy and complete this general equation

__________ + fatty acid → ester + water

Diagram B shows the general formula for glyceryl tristearate: an ester made from glycerol (an alcohol) and stearic acid (a fatty acid).

When this fat reacts with an alkali, such as sodium hydroxide, the fat is broken down into glycerol and sodium stearate:

sodium hydroxide + glyceryl tristearate → sodium stearate + glycerol

Sodium stearate is a typical soap. It can act as a detergent because it contains the stearate ion (hydrophilic) and a long chain of carbon atoms (hydrophobic).

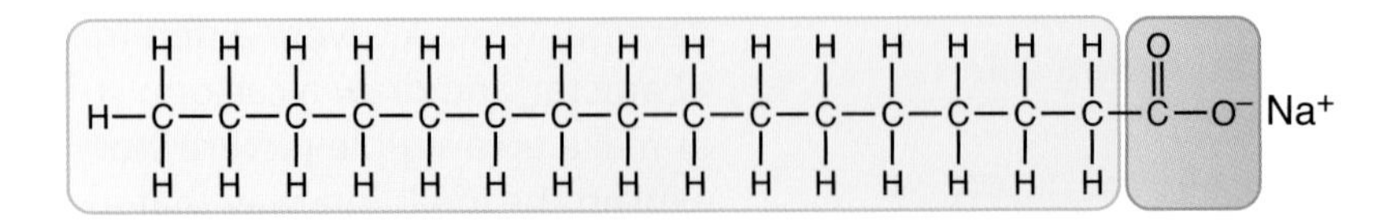

C Sodium stearate: a typical soap.

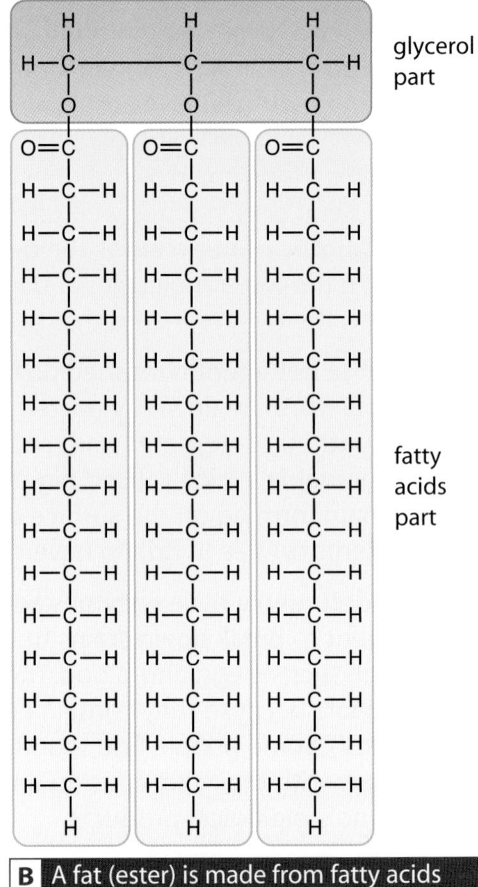

B A fat (ester) is made from fatty acids and glycerol.

5 Copy the diagram of sodium stearate and annotate it to show the hydrophilic and hydrophobic ends.

6 Suggest the name of an oil that might have been used to make the brand of soap shown in photograph D.

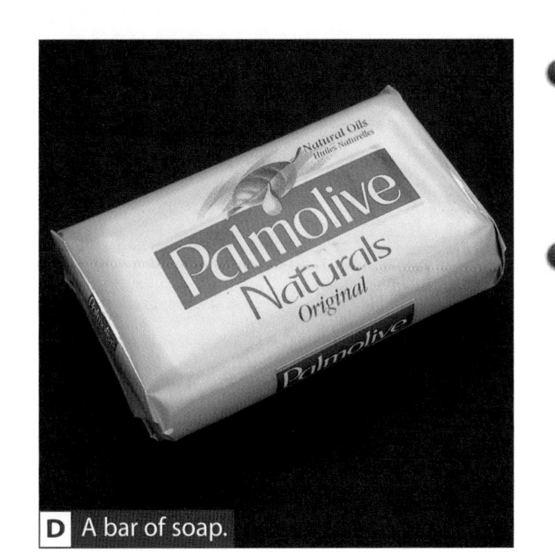

D A bar of soap.

Have you ever wondered?

What is in toothpaste?

Some oven cleaners contain a strong alkali such as sodium hydroxide. The alkali will react with the grease that can get baked on to the oven. When it reacts, it turns the grease into soap, which helps the cleaning process even more. Strong alkalis are corrosive. If you get them on your skin it will feel soapy. This is because the alkalis will start to react with the natural oils in your skin, turning them into soap! Oven cleaners will usually contain a warning to wear gloves to protect your skin from attack.

E Most soaps have a pH that is higher than 7.

Soap itself is usually alkaline. Some people find that soap dries their skin out too much. Recently, a number of products have been developed which have a pH that is closer to 5.5, which is the natural pH value of the skin.

7 Explain why strong alkalis are used in oven cleaners.

8 **a** Estimate the pH of the soaps shown in the photo.
b What is the pH of your skin?
c Is this acid, alkali or neutral?

9 Summarise the similarities and differences between soap and non-soapy detergents.

Higher Questions

16. Hard and soft water

By the end of these two pages you should be able to:

- describe the advantages of using detergents instead of soap in hard water areas
- describe how sodium carbonate is used to produce soda crystals.

A The reaction of soap with soft and hard water.

Soft water gives a good lather when you use soap. With **hard water** it is much more difficult to get a lather. Instead the soap reacts with the water, forming a white precipitate known as scum. This can make it much harder to clean the sink or the bath after you have used it. Products like shower gel and shampoo containing sulphate detergents made from sulphuric acid work equally well in hard and soft water. They do not form a scum.

1 Describe two differences between the reaction of soap with soft and hard water.

2 State one advantage of non-soapy detergents over soap.

Hard water contains minerals (especially calcium and magnesium) which react with the soap. For example:

calcium sulphate + sodium stearate
$\rightarrow$ calcium stearate + sodium sulphate

It is the calcium stearate that is insoluble and makes the scum in the water. Although this is inconvenient for washing, hard water is better in other ways. The dissolved calcium is good for your teeth and bones.

3 State one advantage of hard water over soft water.

4 Write a word equation for the reaction of magnesium sulphate with soap.

Limestone areas of the country usually have hard water. Acids in rain water react with the calcium carbonate in limestone to form soluble calcium hydrogen carbonate. This washes into water sources such as rivers and lakes.

calcium carbonate + water + carbon dioxide
→ calcium hydrogen carbonate

B Limescale in a kettle.

The calcium hydrogen carbonate not only makes the water hard, but can lead to another problem. When the water is heated up, the reaction reverses, and solid calcium carbonate can be deposited as a solid limescale or 'fur' in kettles and pipes. In limestone caves, this is the process that produces stalactites and stalagmites – as the water drips from the ceiling, some of it can evaporate and leave behind the solid calcium carbonate.

5 a What is the main chemical in limestone?
b What chemical is formed when limestone reacts with rain water?

6 What happens when a kettle gets 'furred up'? Write a word equation for the reaction that takes place.

To soften hard water, you have to remove the calcium. Washing soda crystals contain sodium carbonate. Soda crystals get rid of the calcium by turning it into a solid which falls out of the water as a precipitate.

$Na_2CO_3(aq) + CaSO_4(aq) \rightarrow CaCO_3(s) + Na_2SO_4(aq)$

Dishwashers contain a compound which will swap the calcium ions in the water for sodium ions, keeping the water soft. From time to time, you need to 'recharge' the compound by adding salt – this gets rid of the calcium and replaces it with sodium again. Some water softeners contain a compound which traps the calcium in a way which means it will not react with the soap, but it does not form a solid deposit either. In washing machines this reduces the amount of detergent needed and stops limescale forming on the heating element.

C A water softener.

7 How does the formula equation show that the washing soda has removed the calcium from the water?

8 Why is 'Calgon'® a good name for a water softener?

9 Is the water where you live soft or hard? Do you think this is a good thing? Make a list of the advantages and disadvantages – would you prefer a different type of water?

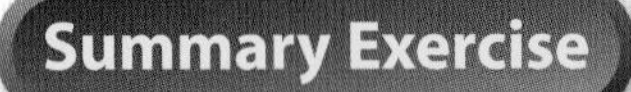

Higher Questions

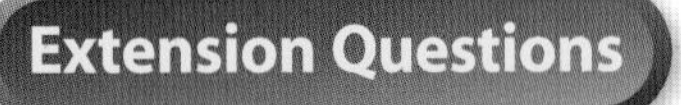

17. Questions

D

P

1 Here are a number of uses of metals. In each case, state one property of the metal that makes it suitable for that use:
  a copper in electrical wiring
  b brass for door handles
  c bronze for coins and medals
  d brass for bells
  e stainless steel for cutlery
  f gold for jewellery
  g aluminium for saucepans
  h iron for making weights for bodybuilders.

2 Graphite is a form of carbon which has a very high melting point. It conducts electricity well, is a very dark grey colour and can be slightly shiny, although you can't see your reflection in it. It has a low density and shatters when you hit it with a hammer. Would you classify graphite as a metal or a non-metal? Explain your reasoning.

3 Choose one of the substances from the following list that is:
  a used in paints or pigments
  b used as a catalyst in the Contact process
  c used as a raw material in the Contact process
  d purified by electrolysis
  e coated on to other metals by electroplating
  f used in stained glass.

  Each substance should be used once only.

  Substances: chromium, copper, copper oxide, iron sulphide, titanium dioxide, vanadium oxide.

4 The manufacture of sulphuric acid uses large amounts of sulphur or iron sulphide, both of which have to be imported in to the UK by ship. The manufacture of sodium hydroxide uses salt water and large quantities of electricity. The finished products will need to be transported around the country when they have been manufactured.
  a Some of the UK's main chemical factories are situated at Avonmouth, near Bristol, on the Wirral, near Liverpool and in the north east of England near Middlesborough. Explain why these sites have been chosen.
  b Would it be a good idea to build a new chemical factory to make sulphuric acid or sodium hydroxide in the area where you live? Give reasoned arguments to support your answer.

5 A brand of smoothing body exfoliator contains the following ingredients: aqua, sodium laureth sulphate, poly(ethene), silicon dioxide, glycerol, propylene glycol, sodium chloride, sodium hydroxide, citric acid, sorbitol, benzyl benzoate, magnesium chloride, magnesium nitrate.
  Which of these ingredients is:
  a water
  b common salt
  c sand
  d an alkali
  e an organic acid
  f a detergent made from sulphuric acid
  g a compound that would react with soap causing scum
  h an alcohol
  i an ester that is made from benzoic acid?

6 The label on a sink-unblocking liquid says: 'CORROSIVE. Contains sodium hydroxide and sodium hypochlorite. WARNING: Do not mix with any acidic products as dangerous gases (chlorine) may be released. Contains non-ionic surfactants and soap.'
  a Which compound is responsible for making the liquid corrosive?
  b Why might this compound be useful for unblocking the drain?
  c If an acid reacts with the sodium hydroxide, what type of reaction would take place?
  d What raw material is used to make sodium hydroxide?
  e Sodium hypochlorite has the formula NaOCl. Copy and complete the word equation to show how sodium hypochlorite is made from chlorine and sodium hydroxide:
  sodium ____________ + chlorine → sodium hypochlorite + ____________ chloride + water
  f Complete and balance the symbol equation for the reaction:
  $NaOH + Cl_2 \rightarrow$ ________ + NaCl + ____
  g Explain the meaning of 'non-ionic surfactant'.
  h Soap is **not** a non-ionic surfactant. Why not?

7 A car battery is also known as a lead-acid accumulator. It contains a number of cells with terminals made of lead (negative) and lead dioxide (positive), dipped in dilute sulphuric acid. Each cell produces a voltage of 2 V.

a Which substance is the electrolyte?

b If the battery has a voltage of 12 V, how many cells are there in the battery?

c At the negative terminal lead atoms change into lead ions (Pb^{2+}). Is this reaction an oxidation or a reduction? Explain your answer.

d Write an ionic half-equation for the reaction in which lead atoms change into lead ions.

The lead dioxide also changes into lead (II) ions like this:

$PbO_2(s) + 4H^+(aq) + 2e^- \rightarrow Pb^{2+}(aq) + 2H_2O(l)$

e Explain why the lead dioxide has been reduced.

The battery is rechargeable. When the car is moving along, some of the energy from the engine is used to drive a dynamo which re-charges the battery by reversing the chemical reactions in the cell.

f Complete the following energy transfer:

chemical energy in fuel and air →

____________ energy (in engine and dynamo)

→ _________ energy (in wires) →

____________ energy (stored in battery)

g Write equations for the two reactions that take place when electricity is passed through the cell to charge it again.

18. Glossary

D

***alcohol** Carbon compound which contains one or more hydroxyl (–OH) groups.

***alkali** A solution which contains excess OH^- ions, and has a pH greater than 7.

***alkali metals** Metals in group 1 of the Periodic Table which react with water to produce metal hydroxides (alkalis).

anode A positive electrode.

battery A number of electrical cells joined together.

carcinogenic Causes cancer.

***catalyst** A chemical which speeds up a reaction but does not get used up.

cathode A negative electrode.

cell A store of chemical energy that can be used to produce electricity.

***Contact process** The industrial process used to make sulphuric acid.

***cosmetic** A substance which is used to improve the appearance of something (usually of the face or body).

density The mass of a substance divided by the volume.

***detergent** A substance which cleans.

ductile Can be drawn out into thin wires.

***dye** A soluble coloured compound which will form a chemical bond with a fabric.

***electrode** A rod made of metal or carbon which carries the current into the electrolyte.

***electrolysis** Splitting up a chemical using electrical energy.

***electrolyte** A liquid that can be split up during electrolysis.

***ester** Carbon compound made from the reaction of an organic acid with an alcohol.

***fatty acid** An organic acid that contains a long chain of carbon and hydrogen atoms.

***flavouring** Substance added to food to improve the flavour.

functional group A group of atoms in an organic compound which have a particular set of properties, e.g. a hydroxyl group or an ester group.

***hard water** Water which contains calcium or magnesium and does not easily form a lather with soap.

homologous series A series of organic compounds which contain the same functional group but have a different carbon chain length.

hydrophilic Something that is attracted to water.

hydrophobic Something that repels water.

***immiscible** Two liquids which do not mix together and will usually form separate layers, one on top of the other.

inert Something which does not react.

***ion** A particle, atom, or group of atoms which has a positive or negative charge.

ionic A substance that is made of ions, or a bond that is formed when oppositely charged ions attract.

malleable Bendy; can be shaped when hit (e.g. with a hammer).

***miscible** Two liquids which will mix together.

mordant A chemical which binds a dye to a fabric.

***organic acid** A weak acid that contains carbon, and is often derived from natural products.

***oxidation** A reaction in which a chemical reacts with oxygen, or takes in oxygen from another chemical. Also when an atom loses one or more electrons.

***pH** Scale from 1 to 14 used to measure acidity.

***pigment** An insoluble, coloured compound used in paints.

rechargeable A cell or battery that can be re-used by using electricity to reverse the chemical reaction and store the energy again.

***reduction** A reaction in which oxygen is taken away from a chemical. Also when an atom or ion gains one or more electrons.

***soap** A detergent that is made by reacting fats or oils with alkalis.

soft water Water which forms a good lather with soap.

***solvent** A liquid that is used to dissolve another substance (the solute) which is often a solid.

sonorous Makes a ringing sound when it is hit.

***surface tension** The apparent 'skin' of the surface of water which is caused by the force of attraction between the water molecules.

***surfactant** A substance, eg a detergent, which reduces the surface tension of water or another solvent.

***transition metal** A metal that is found in the centre block of the Periodic Table. This block includes common metals such as iron and copper.

*glossary words from the specification

P3.5

Particles in action

A The Large Hadron Collider is buried in a tunnel 100 m underground at CERN.

B Playing with a plasma ball at the CERN visitor centre.

Perhaps the biggest experiment in the history of science is the Large Hadron Collider (LHC) at CERN (the European particle physics laboratory) near Geneva. The LHC is a tunnel 27 km long, surrounded by various particle detectors. Superconducting magnets pull protons or heavy ions around the tunnel in opposite directions at almost the speed of light. They collide head-on, producing some completely new types of particles. By studying the particles produced in the LHC, scientists are unlocking secrets of the universe.

This is just the latest in a spectacular series of particle discoveries that have brought us spin-offs such as diesel engines, radiotherapy, televisions and more recently the Internet.

In this topic you will learn that:

- gases are affected by temperature and pressure
- unstable isotopes and their emissions may be identified by the position of the isotope on a neutron/proton curve
- beams of electrons may be produced by an electron gun and carry energy that may be converted into X-rays
- electron beams are used in a variety of equipment including televisions and oscilloscopes.

Say whether you think each of the following statements is true or false.

- Atoms are the building blocks for molecules.
- The particles in a gas are far apart with a high kinetic energy.
- Gas pressure is due to particles hitting surfaces.
- An atom has electrons orbiting its nucleus.
- Opposite charges attract. Like charges repel.
- An ion is an atom that has gained or lost an electron.
- An electric current is a flow of electrons.

1. Absolute zero

By the end of these two pages you should be able to:

- describe the term 'absolute zero'
- convert between kelvin and Celsius scales of temperature
- describe how the temperature of a gas relates to the speed and kinetic energy of its particles
- recognise that the kelvin temperature of a gas and the average kinetic energy of its particles are directly proportional.

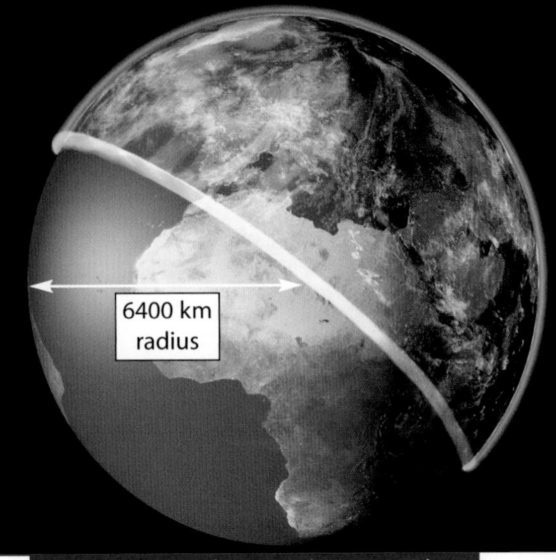

A The thickness of the Earth's atmosphere is shown by the blue lines. At 100 km up you are almost in outer space. Yet half the air lies in the first 5 km and the total volume of this gas could fit into a sphere just 2000 km across!

Imagine being able to take a photograph of the **particles** in the atmosphere. They are continually colliding with each other and their surroundings. Just like anything else that moves they have **kinetic energy**. Raise the **temperature** and they move faster and faster, and knock each other further apart. Lower the temperature and they slow down, coming to a stop at –273 °C when they have no kinetic energy. The temperature –273 °C is called **absolute zero** or zero **kelvin** (0 K for short).

To convert a temperature in degrees **Celsius** into kelvin, you need to add on 273. For example, 20 °C is the same as 293 K.

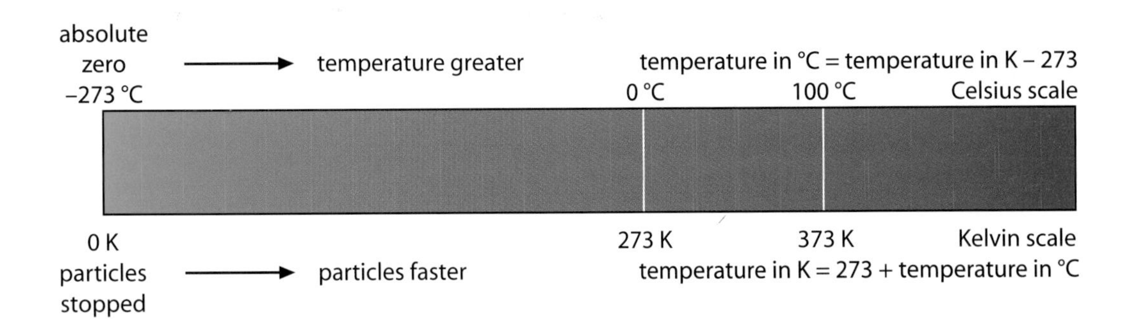

B The Celsius and kelvin scales of temperature.

1 **a** What is 273 K in degrees Celsius?
b What is 27 °C on the kelvin scale?

2 What can you say about the size of degree intervals on the kelvin scale and the size of degree intervals on the Celsius scale?

If you double the kelvin temperature, the average kinetic energy of the particles in a gas also doubles. The kelvin temperature and the average kinetic energy of the particles increase in direct proportion. Therefore you get a straight-line graph through the origin if you plot the kelvin temperature against the average kinetic energy of the particles.

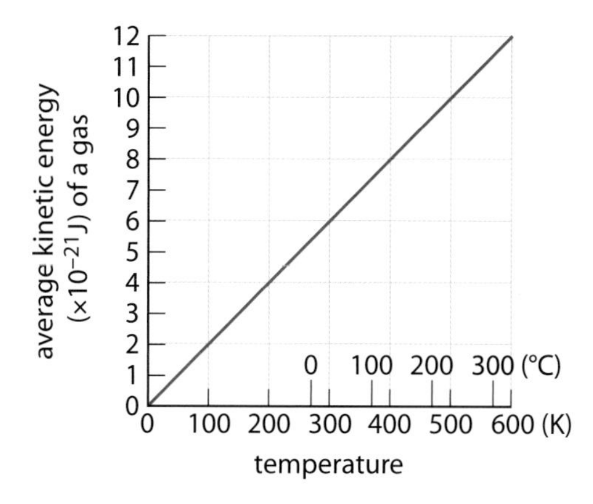

C The temperature of a gas and the average kinetic energy of its particles are directly proportional.

3 a The average kinetic energy of air molecules at 300 K is 6×10^{-21} J. What is the temperature (in K) if their average kinetic energy increases to 12×10^{-21} J?
b What happens to the kelvin temperature?

Suppose you travel from the Earth to the outer planets – the further the planet is from the Sun, the cooler it is. By the time you reach Uranus the temperature is less than –200 °C. The liquid-in-glass thermometers you use in laboratories no longer work. The liquid is frozen. However, a thermometer filled with helium gas might do the job, as helium remains a gas down to –269 °C.

4 a What is –269 °C on the kelvin scale?
b A helium thermometer is shown on the left of diagram D. Describe how it works.

5 A student warms the air in the flask shown on the right of diagram D.
a Explain why the water level falls in the connecting tube.
b Suggest a use this arrangement could be put to.

6 Describe how temperature affects the kinetic energy of the molecules in the air, and why raising the temperature usually causes a gas to expand.

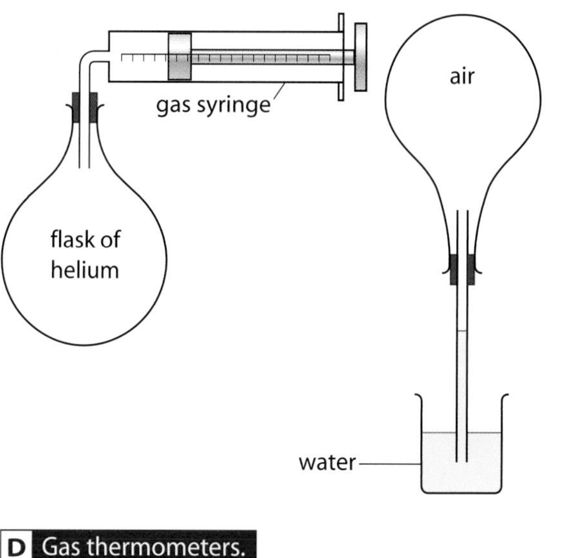

D Gas thermometers.

Summary Exercise

Higher Questions

2. Pressurising particles

By the end of these two pages you should be able to:

- explain how the pressure of a gas depends on the motion of its particles
- use the equation P/T = constant for a gas in a sealed container.

Hold your fingers about one millimetre apart. At the Earth's surface there are roughly one million air molecules between your fingers. If you climb up a mountain, the particles in the air are more spread out. In the vacuum of outer space you would be lucky to find any particles at all. On the Earth the particles continually collide into you and exert an atmospheric **pressure**. In outer space there is none of this bombardment, so there is no pressure.

The atmospheric pressure you normally experience is about 10 N/cm^2 (or 100 kN/m^2). This means that the air is squashing you in with a force of 10 N on every square centimetre of your body. Fortunately the collisions of molecules inside your body balance this atmospheric pressure.

Increasing the temperature increases the effect of the molecules' collisions. At a higher temperature, molecules collide harder and more often. At higher temperatures there is a higher pressure.

1 How do air molecules exert an atmospheric pressure?

2 What is the normal atmospheric pressure in kPa (kilopascals)? (1 kPa = 1 kN/m^2)

3 The molecules inside a tyre heat up on a journey. At a higher temperature what happens to:
a the motion of the molecules
b the force of each collision
c the number of molecules colliding against the tyre walls each second
d the pressure the molecules exert?

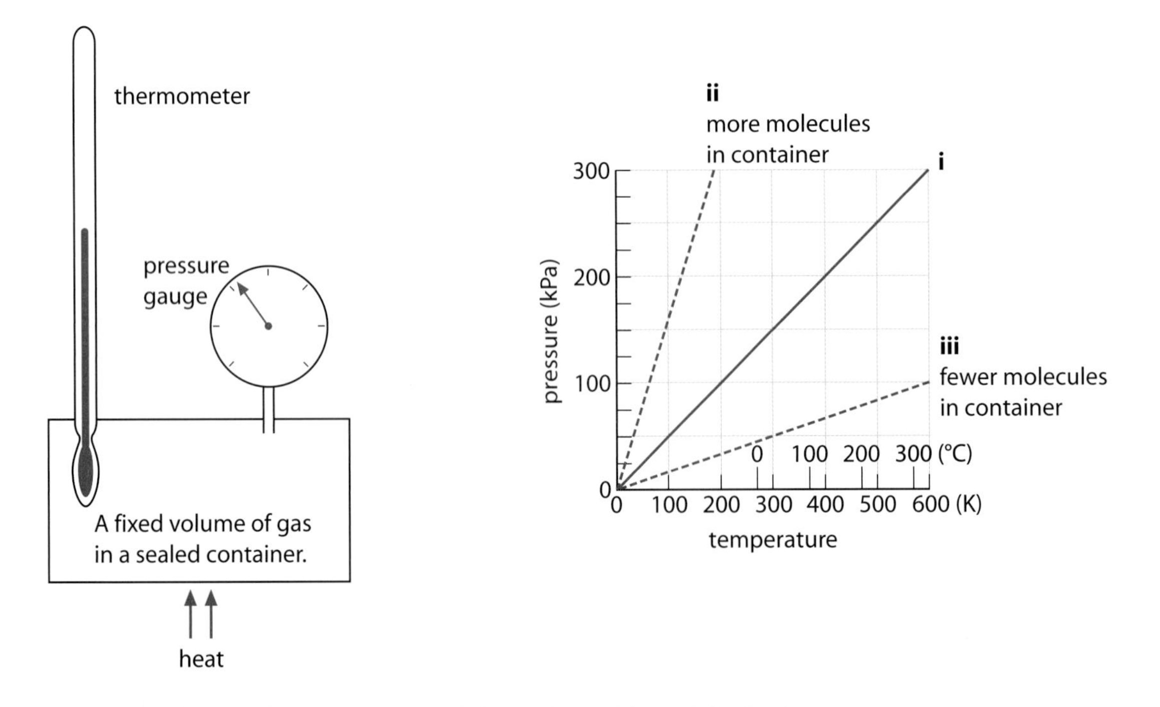

A Apparatus used to test the pressure of a gas at different temperatures.

The equation for a gas in a sealed container is:

$$\frac{\text{pressure, } P \text{ (Pa)}}{\text{temperature, } T \text{ (K)}} = \text{constant}$$

4 a Look at graph **i** on diagram **A**. When the temperature is 300 K the pressure is 150 kPa. What is the pressure at 600 K?

b Divide the pressure (P) by the temperature (T) for the following pairs of readings: 150 kPa, 300 K and 300 kPa, 600 K. What do you notice?

5 Before a match, squash players hit the ball repeatedly against the wall of the court.

a Explain in terms of particles how this affects:

i the temperature of the ball

ii the internal air pressure.

b The pressure inside the ball is 103 kPa at 20 °C. What is the pressure inside the ball after warm-up when the temperature rises to 45 °C?

B Thermogram of a Porsche – the pink areas show high temperatures. The pressure inside its tyres increases as their temperature increases.

There is more than one way to boil water. You can heat it or lower the pressure above it.

You normally supply energy as heat to boil water. Unless you give the water molecules enough kinetic energy they get knocked back down into the water by the air molecules above the surface. Increase the temperature and they escape.

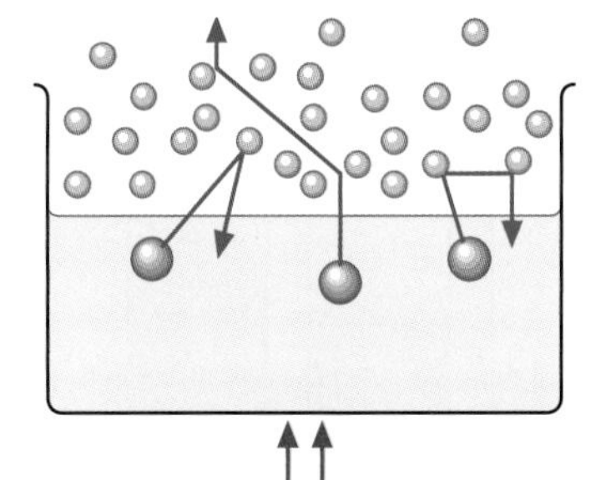

Hot water molecules have more energy, so they can escape.

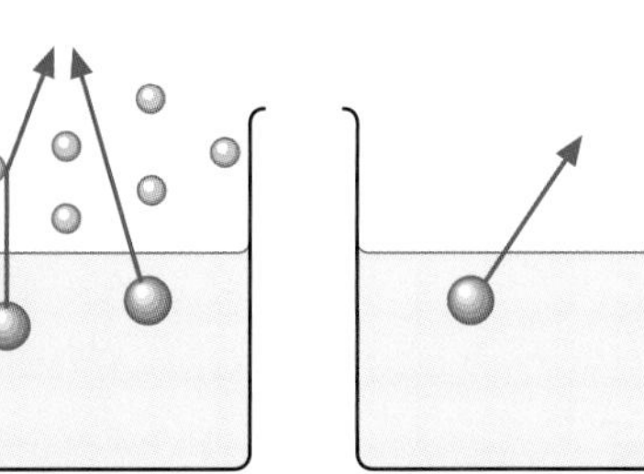

At the top of a mountain, atmospheric pressure is lower. There are fewer air molecules, so water molecules can escape.

In space there are no air molecules, so water molecules can escape easily.

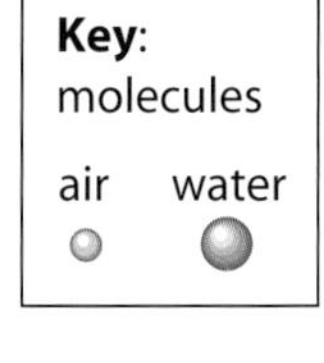

C Three ways to boil water.

However if you go up a mountain, the molecules are further apart, the atmospheric pressure is less, and the water molecules escape more easily. So, at the top of a mountain water boils at *less* than 100 °C. In outer space there is nothing to stop water boiling at a very low temperature, without you even heating it!

Have you ever wondered?

Would our bodies explode if we went into space without a space suit?

6 Describe why the pressure decreases as you climb a mountain. Think about temperature and altitude and write about particle collisions.

3. Gas laws

By the end of these two pages you should be able to:

- use the equation $P_1V_1/T_1 = P_2V_2/T_2$ to explain how gases are affected by temperature and pressure.

The volume of air in a balloon depends on the number of particles it contains, the temperature, and the pressure exerted by collisions from molecules in the surrounding air. A weather balloon can be used to take measurements as it rises through the atmosphere. As a weather balloon gains altitude, the air pressure drops more than the temperature decreases, so the balloon expands. At about 35 km, the balloon expands so much it bursts, returning the instruments to Earth by parachute, with a record of the weather conditions at different altitudes.

A Launching a weather balloon.

1 a As the atmospheric pressure decreases, what happens to the collision rate from molecules on the outside of a weather balloon?

b When the balloon gains altitude, why do collisions from particles on the inside of the balloon cause it to blow up?

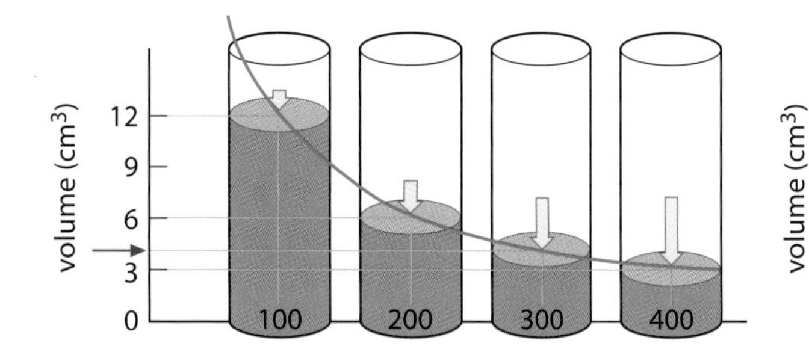

pressure (kPa)

Compressing a gas at constant temperature.
(Boyle's Law)

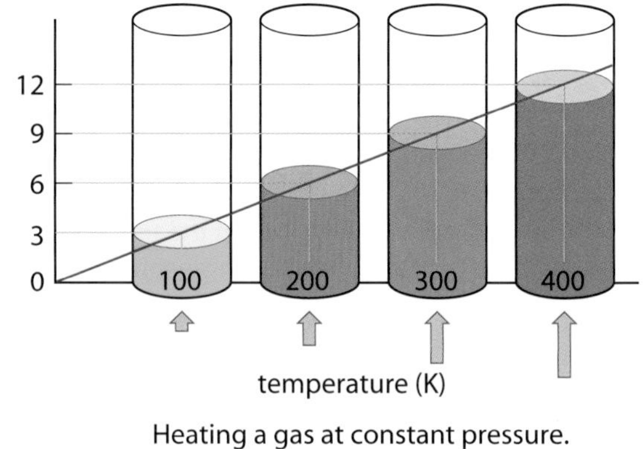

temperature (K)

Heating a gas at constant pressure.
(Charles' Law)

B Compressing and heating a gas.

The graphs show that varying the pressure or the temperature of a gas changes its volume. The first graph shows that volume and pressure are inversely proportional. If the pressure doubles, the volume halves (assuming the temperature remains the same). The second graph shows that volume is proportional to kelvin temperature. If the kelvin temperature doubles the volume doubles, as long as the pressure does not change.

2 In the graph on the left of diagram B, what is the volume of gas at a pressure of 300 kPa?

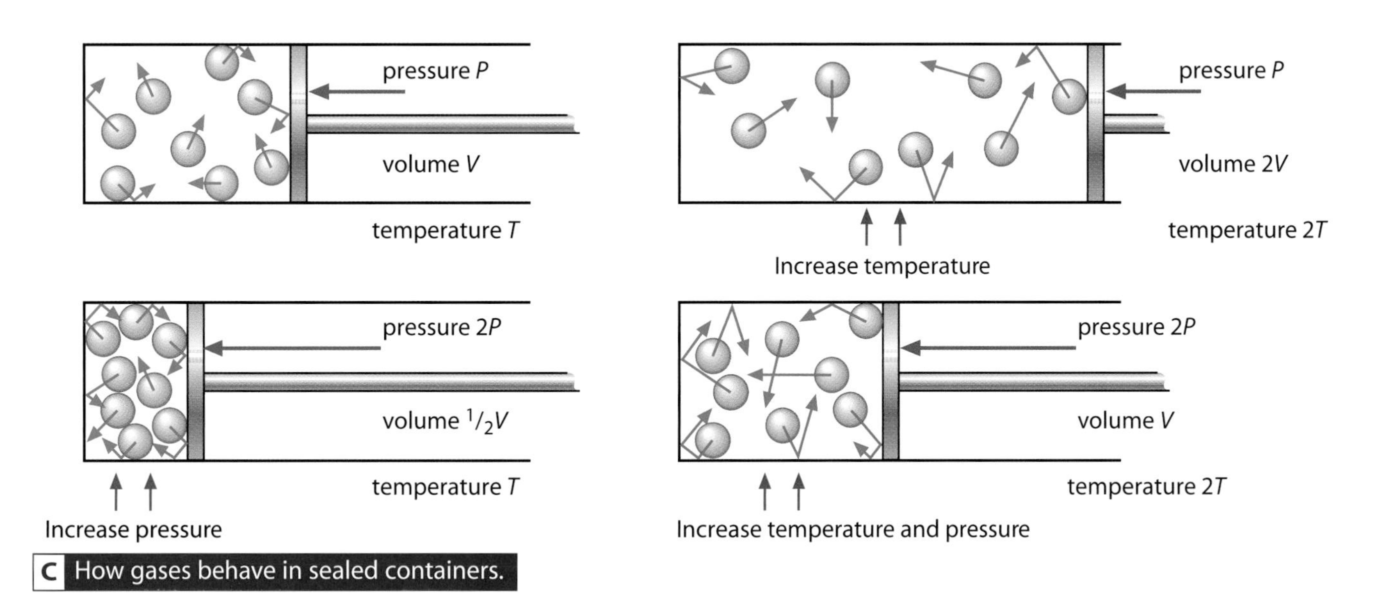

C How gases behave in sealed containers.

The volume of a gas in a sealed container is:

- proportional to the kelvin temperature – if T doubles, V doubles.
- inversely proportional to the pressure – if P doubles, V halves.

The pattern is summarised by the gas equation: $\frac{PV}{T} = \text{constant}$

Imagine the gas trapped in a weather balloon. The volume is $2\ m^3$ at ground level where the temperature is 300 K (27 °C) and the pressure is 1.5×10^5 Pa:

$$\frac{P \times V}{T} = \frac{1.5 \times 10^5 \times 2}{300} = 1000$$

At a high altitude, the temperature falls to 240 K (–33 °C) and the pressure drops to 0.3×10^5 Pa. The gas equation helps you to calculate the new volume V:

$$\frac{P \times V}{T} = \frac{0.3 \times 10^5 \times V}{240} = 1000$$

$$125V = 1000$$
$$V = 1000/25 = 8\ m^3$$

The gas equation can be written as:

$$\frac{P_1V_1}{T_1} = \frac{P_2V_2}{T_2}$$

where T is measured in kelvin (not °C). Subscript 1 refers to the starting conditions and subscript 2 refers to the final conditions. For the weather balloon above:

$$\frac{1.5 \times 10^5 \times 2}{300} = \frac{0.3 \times 10^5 \times 8}{240}$$
$$1000 = 1000$$

3 A bicycle pump holds a volume of $100\ cm^3$ when the piston is drawn out. The air is initially at a temperature of 280 K (7 °C) and a pressure of 100 kPa. If compression reduces its volume to $20\ cm^3$ and increases its temperature to 336 K (63 °C), show that the pressure of the air forced into the tyre is 600 kPa. Use

$$\frac{P_1V_1}{T_1} = \frac{P_2V_2}{T_2}$$

4 A $5\ m^3$ helium balloon is released from sea level where the temperature is 20 °C and the pressure is 100 kPa. What is its volume, just before it bursts, when the temperature is –20 °C and the pressure is 5 kPa?

5 The tyres of Formula One racing cars are not inflated by pumping more air into them, but by relying on the increase in temperature that occurs when they heat up in the race. Explain why, discussing the movement of particles.

Summary Exercise

Higher Questions

Extension Questions

4. An electron gun

By the end of these two pages you should be able to:

- describe the process of thermionic emission – electrons 'boiling off' hot metal filaments
- explain how an electron gun, with heated cathode and accelerating anode, produces a beam of electrons (which is equivalent to an electric current)
- explain the factors that affect the deflection of an electron beam or a stream of charged particles by an electric field between charged metal plates.

A neutral atom can be thought of as a positive ion and a negative outer electron. Raise the temperature and **thermionic emission** occurs – the atom is so hot it changes to an ion and emits the **electron**. This happens when an electric current heats up a metal filament – electrons are 'boiled off'. Although this occurs in a light-bulb filament, you are not aware of the electrons being emitted. Unlike the light from the filament, the electrons cannot pass through the glass bulb.

Diagram A shows thermionic emission near an anode (positive plate). The electron is attracted and accelerates towards the anode. If the anode has a hole, the electron can pass through it. This principle is used in electron guns.

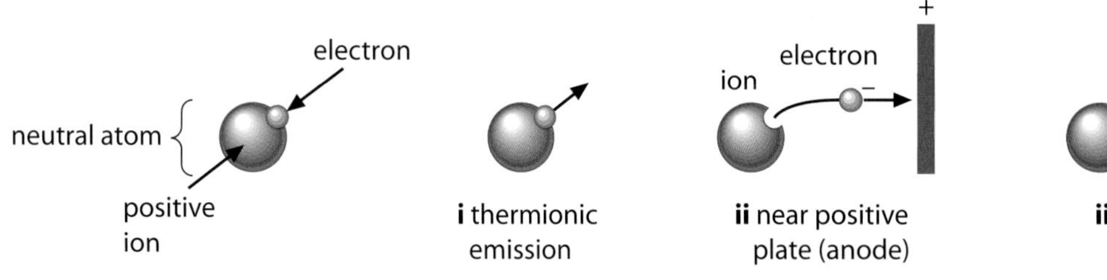

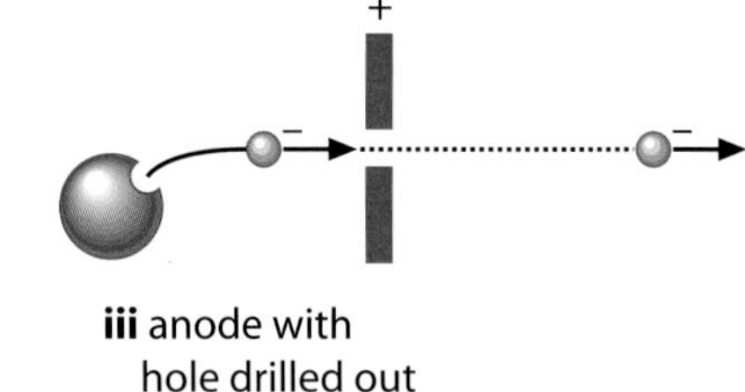

A Thermionic emission near a positive plate.

Experiments by J. J. Thomson in 1897 led to the discovery of the electron, and the award of the 1906 Nobel prize. He detected electrons as they landed on a fluorescent screen. Diagram B shows one of Thomson's electron guns emitting a beam of electrons, which pass between two charged parallel plates. The whole apparatus is in a vacuum, otherwise air molecules would stop the electrons by getting in their way.

The 6 V supply to the filament is sufficient to heat the cathode and emit electrons. A voltage of 4000 V between the **cathode** (negative electrode) and anode (positive electrode) accelerates the electrons to high speeds.

Look at the diagram. Remember unlike charges attract, and larger voltages produce bigger charges.

1 What is thermionic emission?

2 What happens if electrons 'boil off' atoms near a positive plate?

3 What happens if thermionic emission occurs near an anode with a hole in it?

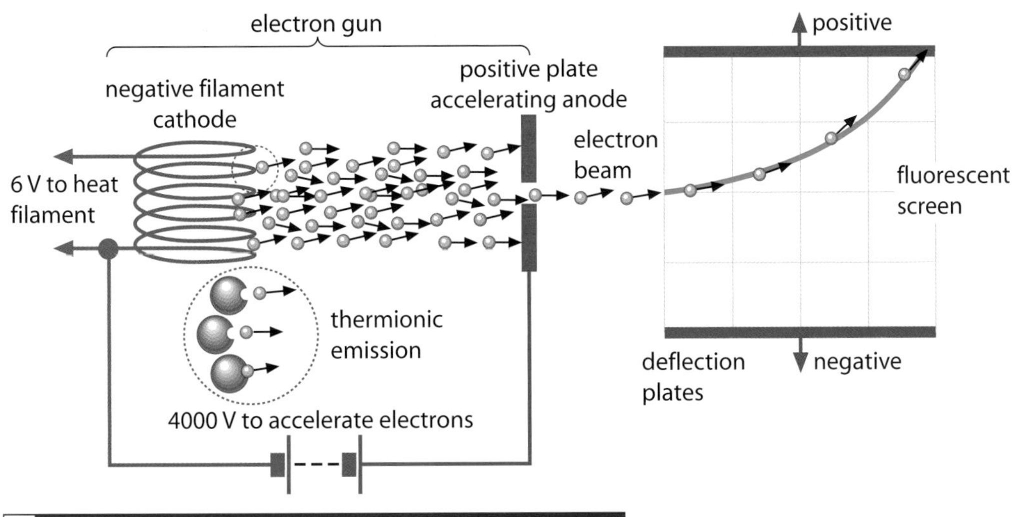

B An electron gun and deflection of an electron beam.

Some electrons pass through the hole and come out as an **electron beam** on the other side. Parallel charged metal plates create an electric field and deflect these electrons. Since the **accelerating anode** gives them enough kinetic energy, the electrons make the fluorescent screen glow and show the path of the beam. Changing the voltage between the anode and cathode, or between the deflection plates, affects the path of the electron beam.

Diagram C shows negatively charged ink drops falling between oppositely charged parallel plates under different conditions. The path of an ink drop bends like an electron beam.

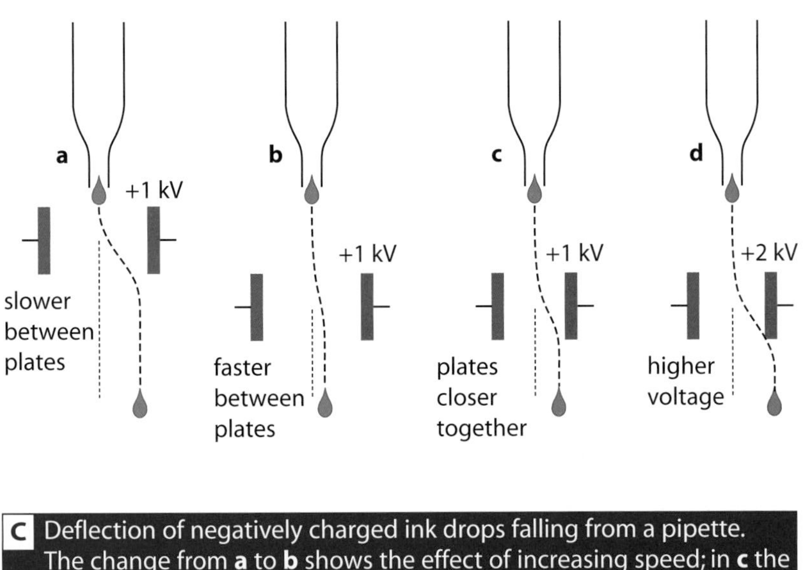

C Deflection of negatively charged ink drops falling from a pipette. The change from **a** to **b** shows the effect of increasing speed; in **c** the distance between the plates is decreased; in **d** a higher voltage is applied to the plates.

4 How does the path on the fluorescent screen prove that the electrons are negative?

5 If you connect a higher voltage to the deflection plates, what happens to the deflection of the beam?

6 If you connect a higher voltage to the accelerating anode:
- **a** what happens to the kinetic energy of the electrons
- **b** how does the path on the fluorescent screen change now?

7 a Look at diagram C. If the metal deflection plates are placed lower down, where the ink drop moves faster, the drop does not spend so much time between the plates. What effect does this have on the deflection?

b If the plates are closer together or their voltage is greater, the electric field is stronger. What effect does this have on the deflection?

8 Use the words 'thermionic emission' and 'accelerating anode' to explain how to produce a beam of electrons using an electron gun.

Higher Questions

5. Oscilloscopes

By the end of these two pages you should be able to:

- describe the use of electron beams in oscilloscopes
- use the equation kinetic energy = electronic charge × accelerating voltage (or KE = $e \times V$).

Oscilloscopes are voltmeters that display how a voltage changes with time. Oscilloscopes can also show real-time graphs of how variables like temperature and pressure change. This makes them useful in hospitals to monitor the condition of patients. First the variable is converted to an electrical signal, then an electron gun in the oscilloscope fires electrons onto the screen to display the image.

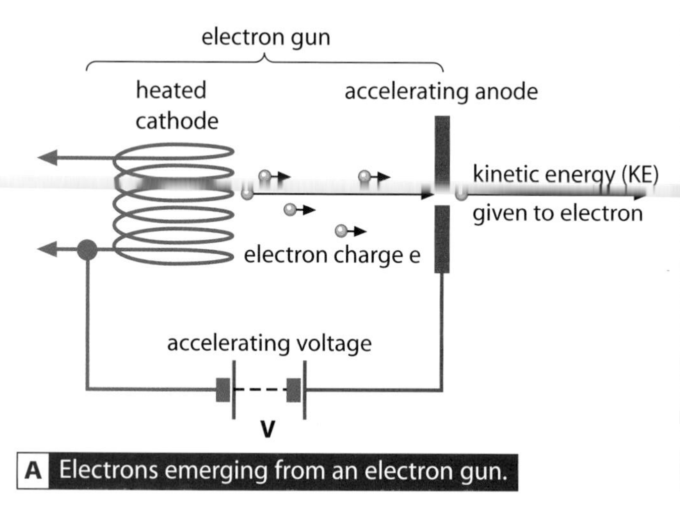

A Electrons emerging from an electron gun.

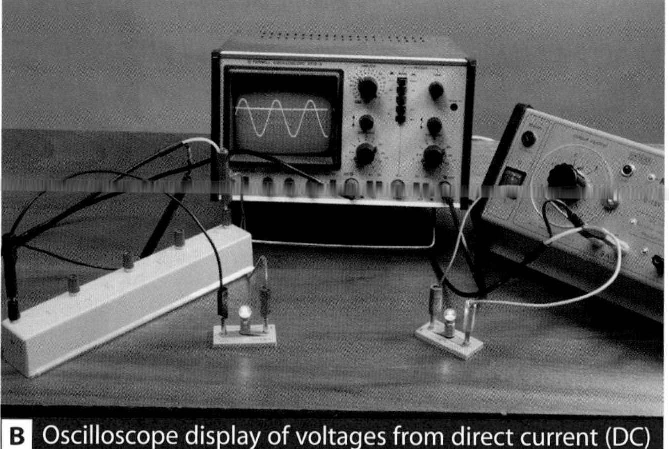

B Oscilloscope display of voltages from direct current (DC) and alternating current (AC) supplies

Scientists sometimes call an electron beam a cathode ray because the electron beam travels from the cathode (the negative electrode) in an electron gun. **Cathode ray tubes** (**CRTs**) are **evacuated** tubes containing one or more electron guns. Before the arrival of flat screen technology, CRTs were used in televisions and computer monitors.

1 What are oscilloscopes used to display?

2 Why do electrons repel away from a cathode?

Look at the diagram of an oscilloscope (diagram C). The three main parts are the electron gun, deflection plates (*x*- and *y*-plates) and fluorescent screen. The voltage applied to the *x*-plates controls the **time base** or how quickly the electron beam moves across the screen. The voltage applied to the *y*-plates affects the pattern displayed on the screen. The voltage amplification control adjusts the size of this display – the greater the *y*-voltage the higher the amplitude of the image.

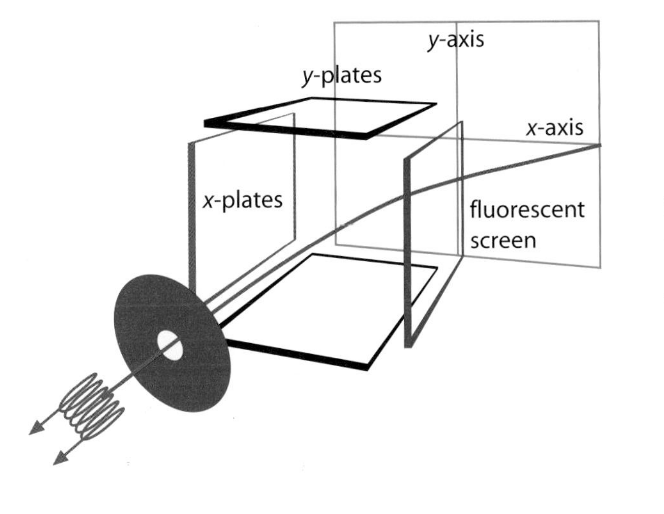

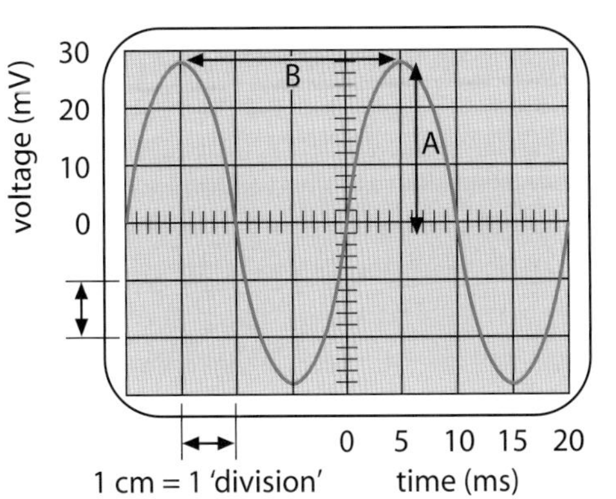

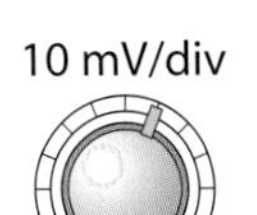

C Inside an oscilloscope and controlling the image on the screen.

3 On the oscilloscope display above, what is:

a the peak voltage (A) of the alternating current in mV (millivolts)?

b the time period (B) from one peak to the next in ms (milliseconds)?

c the frequency at which the current alternates in Hz (ie the number of waves or time periods per second)?

A voltage provides energy to the electrons flowing in an electric circuit. In the same way the voltage between cathode and anode in an electron gun gives kinetic energy to the electrons in an electron beam.

To calculate the kinetic energy gained by one electron fired from an electron gun, you use the equation:

kinetic energy, KE = electronic charge, $e \times$ accelerating voltage, V

where the electronic charge (or charge on one electron), $e = 1.6 \times 10^{-19}$ C (coulombs).

The charge on an electron is small – a six million, million, millionth of a coulomb. Even an accelerating voltage as high as 4 kV does not give an electron much energy:

$\text{KE} = e \times V$
$= 1.6 \times 10^{-19}\ \text{C} \times 4000\ \text{V}$
$= 6.4 \times 10^{-16}\ \text{J}$

Yet, despite the electron's small kinetic energy, because its mass is tiny, the electrons still travel at about one-tenth of the speed of light – the speed of light being 300 million m/s.

4 Write 300 million m/s in standard form.

5 If the accelerating voltage in an electron gun is 3 kV, what is the kinetic energy of the electrons in the beam?

6 Explain how the equation $\text{KE} = e \times V$ applies to an electron gun in an oscilloscope.

Higher Questions

6. Televisions

By the end of these two pages you should be able to:

- describe the use of electron beams in television picture tubes and computer monitors.

An oscilloscope bends an electron beam using charged deflection plates, but in a television tube electromagnets deflect the electrons. The evacuated cathode ray tube (CRT) of a colour television is more complex than that of an oscilloscope. In an oscilloscope the electrons form a single repeating line on the screen to create the image. In a television the electromagnets deflect the electrons in zigzags across and down the screen, making 625 horizontal lines before returning to the start. The electron beam scans the whole screen every 1/25th of a second. The picture consists of three separate dots (or pixels) of colour which your brain assembles into a meaningful picture. To see what is happening look at the blown-up image in photograph A as you move further away.

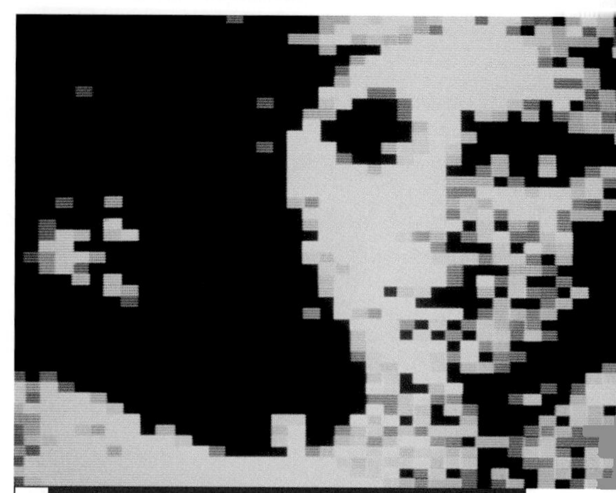

A Each small coloured square on the television screen is called a pixel. Standard videotape is 800 (horizontal) × 600 (vertical) pixels (or 4:3 format).

In colour televisions there are three electron guns, one for each primary colour: red, blue and green. The television screen is coated with dots of three different types of phosphor, one for each colour. A phosphor is a material that emits visible light when exposed to radiation. When electrons hit these phosphors they emit either red, green or blue light.

Before the electron beams hit the phosphors they pass through a shadow mask (see diagram B). The shadow mask ensures that each electron gun only hits one colour-type of phospor. The mixture of light from neighbouring dots gives the illusion of different colours of different intensities. For example, if mostly red phosphors are hit we see red light; if equal amounts of red and green phosphors are hit we see yellow.

To vary the brightness, the number of electrons emitted by the electron guns keeps changing. For example, if there are more electrons, the image is brighter. The beams move rapidly across and down the screen, giving different colours and brightnesses and creating the illusion of a moving picture.

1 Why is a combination of red, green and blue primary colours necessary?

2 How does applying a lower voltage to the accelerating anode of an electron gun affect:

a the speed of the electrons (*Hint*: $KE = e \times V$)

b the number of electrons arriving at the screen

c the brightness of the picture?

3 What is the ratio: number of phosphor dots divided by number of holes in the shadow mask?

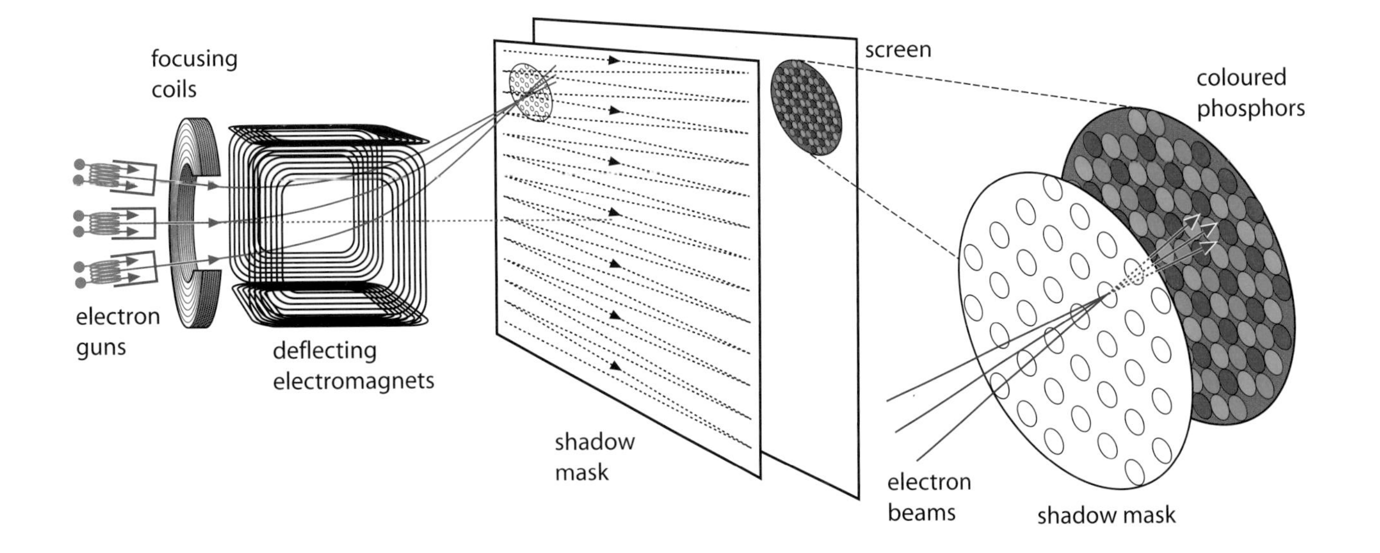

B Inside the cathode ray tube of a television or computer monitor. The enlargement shows the shadow mask for separating the three electron beams, and the screen with red, blue and green phosphors.

Your brain cannot distinguish more than 25 pictures each second, but to avoid flicker the electron beam traces out alternate lines (see diagram C). In the first 1/50th of a second, voltages to the electromagnets move the odd numbered lines across the screen. The even numbered lines scan across in the next 1/50th of second. The graphs show the voltages V_A and V_B applied to the electromagnets. One voltage rapidly moves the electron beam down the screen. The other forces the electrons to scan back and forth across the screen.

4 Which voltage, V_A or V_B, deflects the electron beam
a down the screen
b across the screen?

5 Explain how a colour television works.

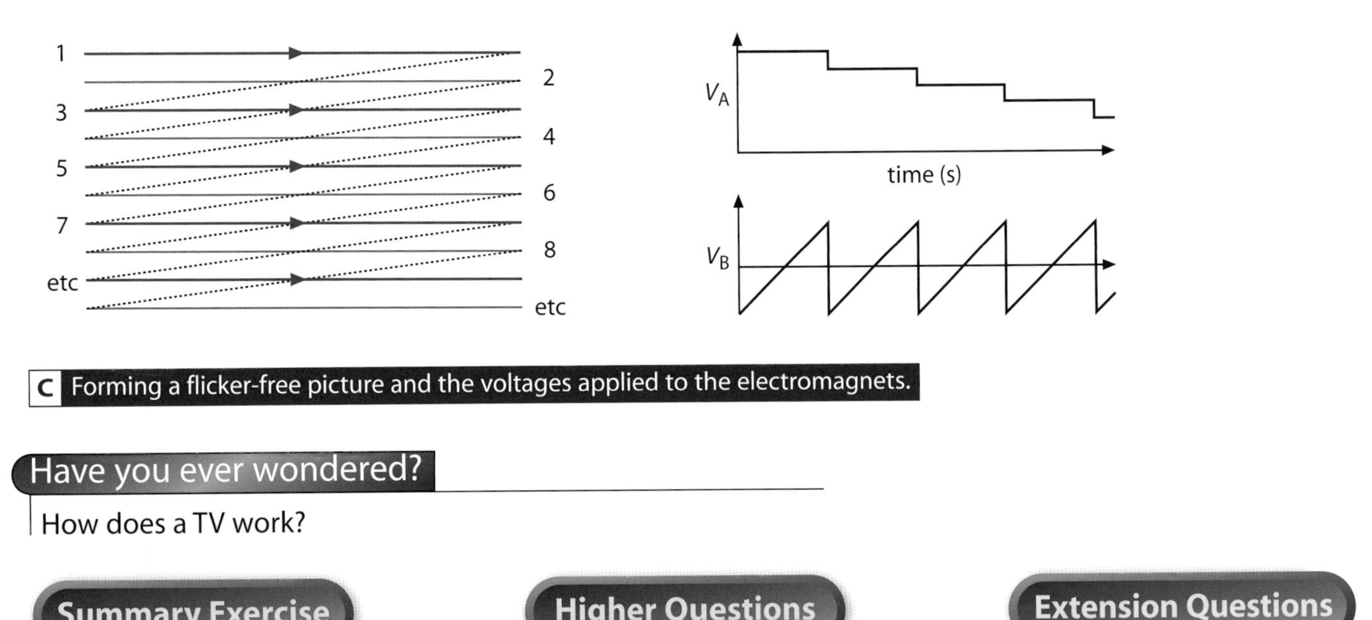

C Forming a flicker-free picture and the voltages applied to the electromagnets.

Have you ever wondered?

How does a TV work?

Summary Exercise

Higher Questions

Extension Questions

7. The production of X-rays

By the end of these two pages you should be able to:

- describe the use of electron beams in the production of X-rays
- describe an electron beam as equivalent to an electric current and calculate the current in terms of the rate of flow of electrons.

An X-ray machine is an evacuated (cathode ray) tube containing a high voltage electron gun. Its high energy electrons collide with a tungsten-alloy accelerating anode. As the electrons crash into the anode they come to a sudden stop. In slowing down rapidly, about one in 100 of these electrons emits an X-ray – which radiates away at a 90° angle. The kinetic energy of the other electrons merely transfers to heat – the anode temperature reaching 2500 °C. Most X-ray machines have a motor which spins the anode. This avoids wear on the same spot and helps conduct the heat away.

1 Tungsten alloy has the highest melting point of all metals – over 3000 °C. Why is tungsten alloy a good choice for the accelerating anode of an X-ray machine?

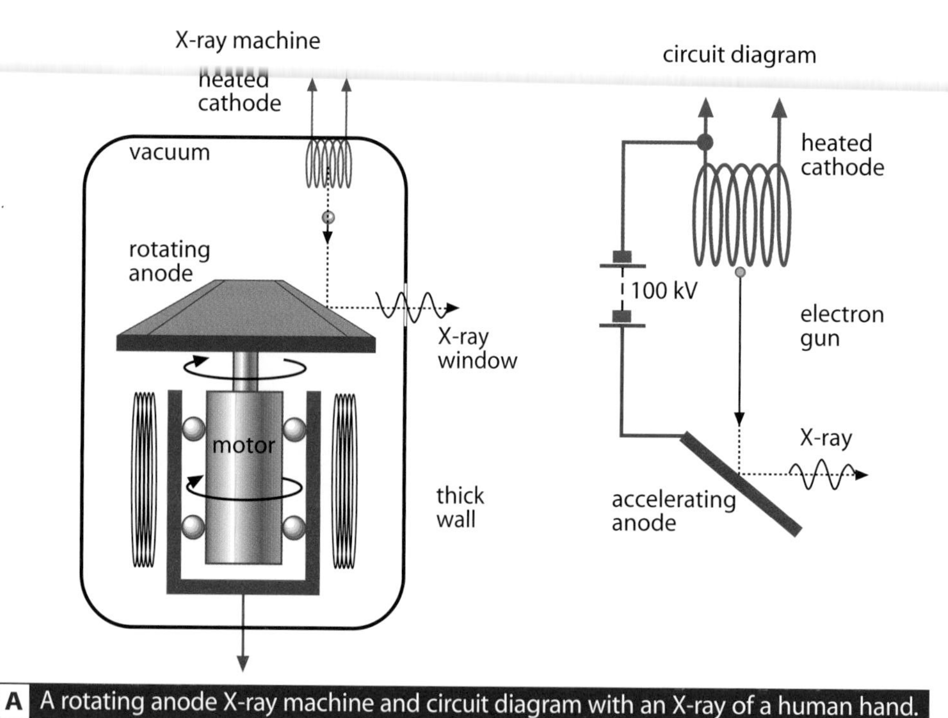

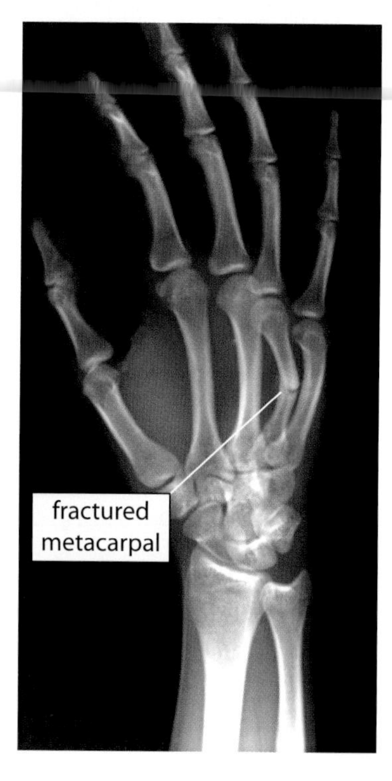

A A rotating anode X-ray machine and circuit diagram with an X-ray of a human hand.

Voltage provides energy to the electrons flowing in both an electric circuit and an electron beam. In both cases electrons leave their atoms.

The current is the rate of flow of electrons.

$$\text{Current } I \text{ (in A)} = \frac{N \times e}{t}$$

where N is the number of electrons flowing in time t and e is the charge on one electron ($e = 1.6 \times 10^{-19}$ C).

2 Show that, in a current of 1 A, there are 6×10^{18} electrons flowing by every second.

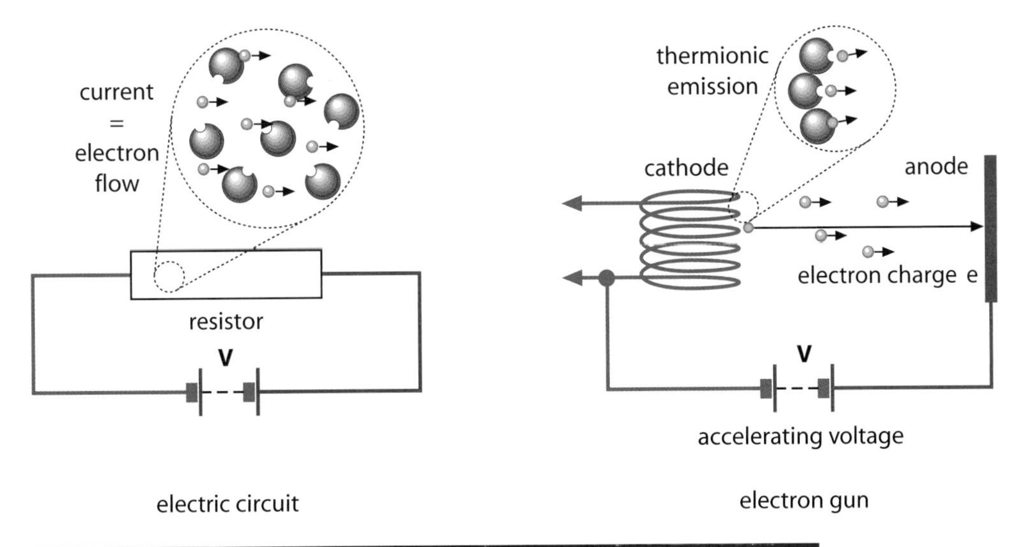

B Comparing the current inside an electric circuit and an electron gun.

A radiographer alters the current and voltage to control the intensity and penetration of the X-ray beam. These changes improve the quality of the image for diagnosis. The radiographer can heat the cathode to emit more electrons each second. This produces a greater electron current and creates more X-rays every second, or a more intense X-ray beam. The radiographer can increase the accelerating voltage applied to the anode, which gives the electrons more kinetic energy. This makes more energetic X-rays with a shorter wavelength. Shorter-wavelength X-rays are more penetrating.

The resulting image you see on photographic film is like a shadow. Dark areas show where X-rays passed straight through your body. These areas are less dense, e.g. fat and muscle. Light areas show dense parts of your body, such as bone, which stop X-rays passing through.

Short-wavelength electromagnetic waves like X-rays damage living tissue. Radiographers must limit the risk to patients by reducing the intensity, exposure time and wavelength of the X-rays, yet obtain a clear enough image for diagnosis. To avoid risk of cancer, radiographers wear lead–rubber aprons and stay behind protective screens when taking X-rays.

X-ray diagnosis is not limited to X-ray photography or the study of broken bones. For example a CT (computer tomography) scan can produce a 3D computer-generated image of soft tissue. In a CT scan, sensors detect low-intensity X-ray pulses as the machine spirals round the patient. A computer processes these multiple shots and displays either 'slices' or 3D images of the tissue. (In Greek, *tomos* means slice, hence the word tomography.)

3 How can a radiographer improve the quality of an X-ray image?

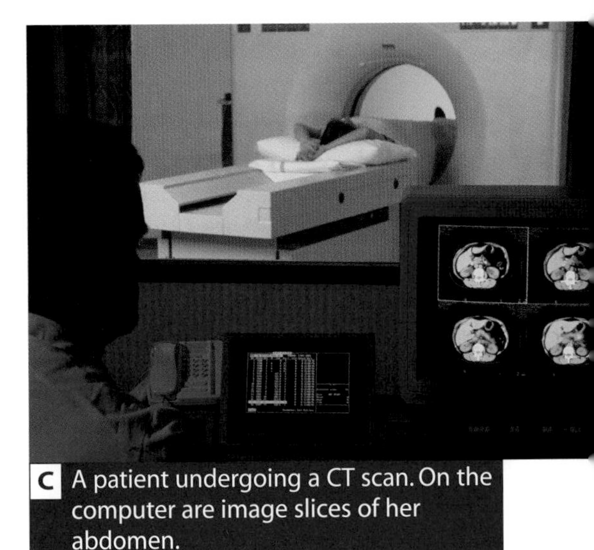

C A patient undergoing a CT scan. On the computer are image slices of her abdomen.

4 The radiation dose from a CT scan is 13 times more than from a single X-ray. What benefit is there with a CT scan to compensate for this additional danger?

5 Explain how an X-ray machine works.

Summary Exercise   Higher Questions   Extension Questions

8. Nuclear radiation

By the end of these two pages you should be able to:

- state that the nuclei of atoms contain protons and neutrons
- explain that neutrons are difficult to detect because they are neutral
- describe the properties of alpha, beta, gamma, positron and neutron radiation.

A beryllium atom has four electrons that orbit its tiny, central **nucleus**. The nucleus consists of four **protons** (red) and five **neutrons** (blue). The number of protons (positive) balances the number of electrons (negative) to make the overall charge of the atom neutral. Each proton and neutron consists of three **quarks**, bound together by a strong nuclear force.

A **radioactive isotope** is an atom with an unstable nucleus. The nuclei of radioactive isotopes can emit different types of **radiation**:

- **alpha particles** (α) – massive, as made of two protons and two neutrons, with charge 2+. Alpha particles are easily stopped by paper or skin.
- **beta particles** (β^-) – high-speed electrons, with charge 1–. Beta particles are stopped by materials with densities like aluminium.
- **gamma radiation** (γ) – pulses of electromagnetic waves with very short wavelengths. Gamma rays penetrate matter easily. They are only stopped by dense lead shielding.
- neutrons (n) – high-speed neutral particles, emitted in nuclear reactions.
- **positrons** (β^+) – the **anti-matter** equivalent of beta particles, with charge 1+.

Anti-matter is made from **anti-particles**. An anti-particle has the same mass as its equivalent particle but opposite properties. For example, a proton is positive, but an anti-proton is negative.

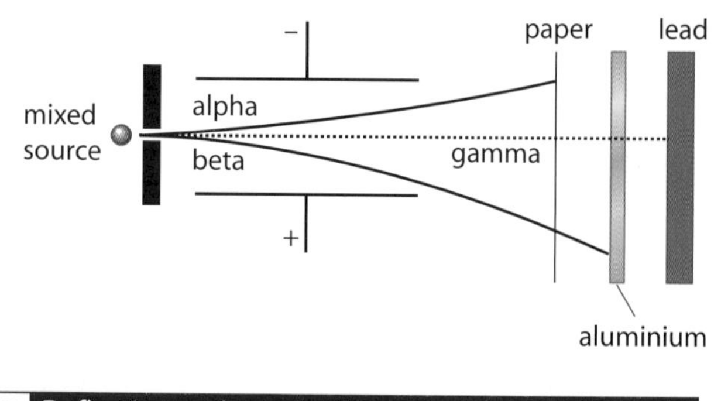

B Deflection and penetration properties of radiations.

A A beryllium atom.

1 What particles does a nucleus contain?

2 Use the information in diagram B to describe the properties of alpha, beta and gamma radiation.

3 Like most particles, electrons are 'matter' particles. What is the 'anti-particle' of an electron having the same mass but a positive charge?

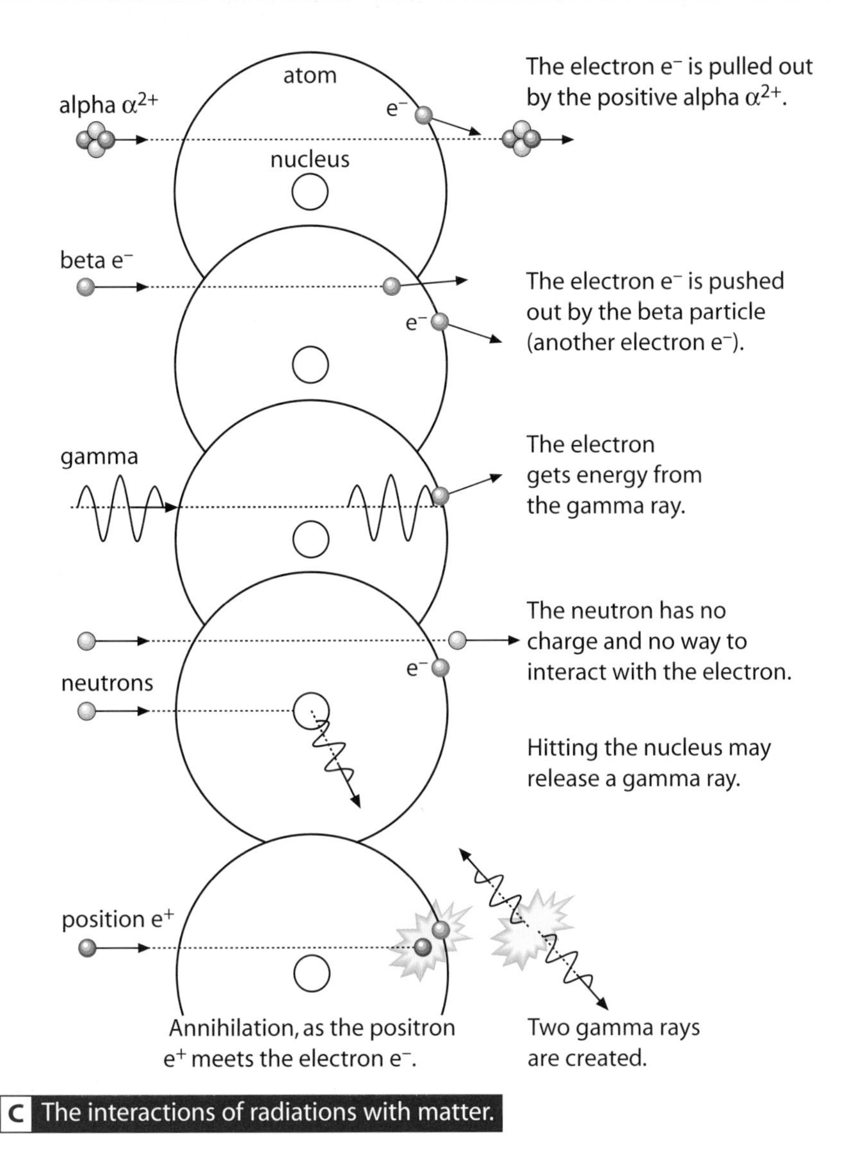

C The interactions of radiations with matter.

Look at diagram C. Alpha, beta and gamma radiations are called ionising radiation because they can eject electrons from atoms. This turns the atoms into ions. Any electron emitted in this way produces a current pulse in a Geiger counter. So you use a Geiger counter to detect ionising radiation.

Neutrons have no charge, therefore have no way to force an electron from an atom. They cannot be detected unless they collide directly with an atom's nucleus. In this case the nucleus may capture the neutron, split in two or emit a gamma ray. However, most of the time the neutron just bounces off the nucleus.

When an anti-particle meets its equivalent particle, they destroy each other and create two gamma rays travelling in opposite directions. Therefore a positron can only be detected when it collides with an electron and creates two gamma rays. This principle is used in positron emission tomography (PET) scans.

4 Alpha particles pull electrons out of atoms, whereas beta particles push electrons out. Explain this difference. (*Hint*: think about the charges of alpha and beta particles.)

5 **a** What normally happens if a neutron collides with a nucleus?
  b Why don't neutrons attract or repel electrons?
  c Why are neutrons hard to detect?

6 What are the five types of nuclear radiation and their properties?

Summary Exercise **Higher Questions**

9. Stable and unstable isotopes

By the end of these two pages you should be able to:

- describe the features of a graph of neutron number (*N*) against proton number (*Z*) for stable isotopes
- use this graph to identify which isotopes of an element are stable and which are radioactive
- state whether an unstable isotope emits alpha, beta or positron particles
- recall that nuclei with more than 82 protons usually emit alpha particles.

Atoms of the same element always contain the same number of protons in their nucleus. However the number of neutrons in their nucleus can differ. Carbon-12 ($^{12}_{6}C$) and carbon-14 ($^{14}_{6}C$) are **isotopes** of the same element. Isotopes are atoms of the same element that contain different numbers of neutrons. Most carbon atoms are carbon-12 and have a stable nucleus (six protons and six neutrons) that does not change. However, carbon-14 atoms are radioactive atoms that have an unstable nucleus (six protons and eight neutrons), which decays and emits beta particles. When the nucleus of a radioactive isotope **decays** it emits either alpha particles, beta particles, positrons or gamma rays.

The chemical symbol for an isotope is written: $^{A}_{Z}X$
where *A* is the mass (or **nucleon**) number and *Z* is the proton (or atomic) number.
mass number A = proton number Z + neutron number N
e.g. carbon-14, $^{14}_{6}C$ has six protons + eight neutrons

Isotopes of the same element have the same proton number. So carbon-13, $^{13}_{6}C$ has six protons and seven neutrons.

Doctors sometimes use radioactive isotopes in medicine. They are given to patients and build up in certain areas, giving off radiation. By detecting the radiation and creating images, the doctors can diagnose the condition of their patients.

Diagram B shows a graph of neutron number against proton number for stable isotopes. Stable nuclei with a lighter mass have about the same number of neutrons as protons (where $N = Z$, near the origin). However in a nucleus with a larger mass, with more protons, the repulsion between the protons is stronger, so these nuclei have more neutrons. This reduces the force of replusion and holds the particles together more firmly. These stable nuclei lie above the $N = Z$ line. Isotopes that do not lie on the **belt of stability** curve have either too few neutrons or too many neutrons. This makes the nucleus unstable and likely to break up.

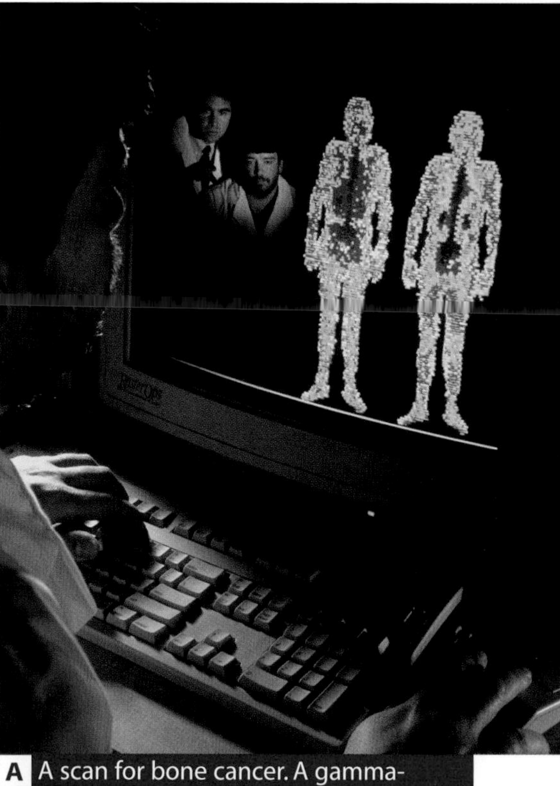

A A scan for bone cancer. A gamma-emitting isotope, technetium-99m is injected into the patient, and goes to cancerous cells. These show as dark areas on the scan.

1 What do all atoms of the same element have in common?

2 Which two of the following are isotopes of the same element? Explain your answer.

$^{15}_{8}A$, $^{14}_{7}B$, $^{16}_{8}C$, $^{14}_{6}D$

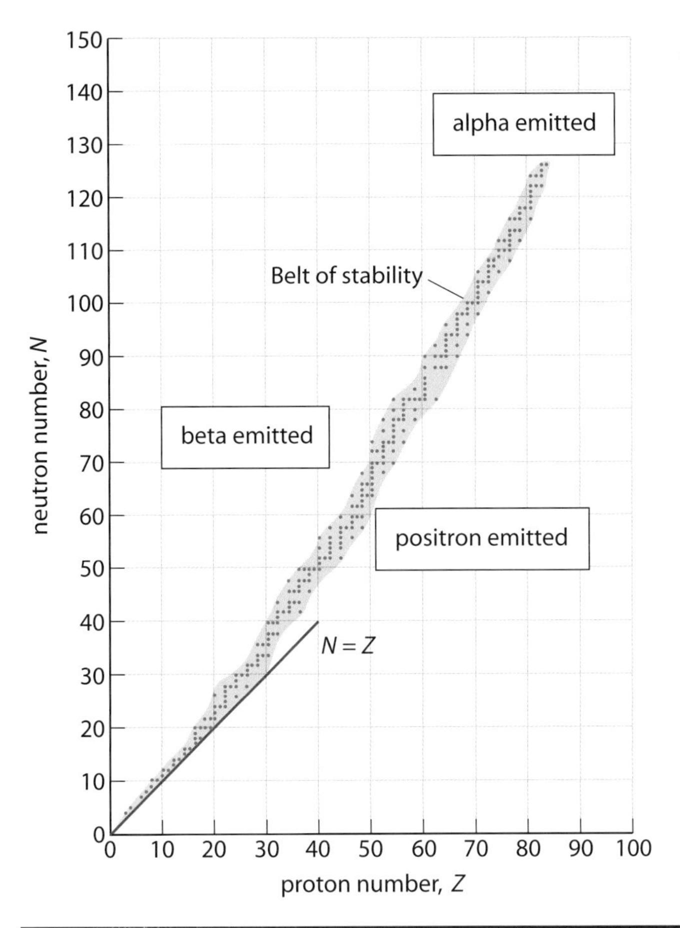

B Graph of N against Z for stable isotopes, showing what type of radiation unstable nuclei emit.

Isotopes that lie below the curve have too few neutrons to be stable. These nuclei are proton/positive-rich and emit a positron (so undergo β^+ decay). Isotopes that lie above the curve have too many neutrons to be stable. These nuclei are not positive enough and emit an electron (so undergo β^- decay).

When β^+ or β^- decay occurs, the remaining protons and neutrons in the nucleus often group themselves differently. Such a rearrangement results in a loss of energy, and gamma radiation is emitted.

All nuclei with a proton number above 82 are unstable. These radioactive isotopes find it easier to emit a larger particle and so emit an alpha particle (two protons and two neutrons).

Have you ever wondered?

How do you find out if something is radioactive?

3 Look at graph B.

a Approximately how many neutrons are there in a stable nucleus containing 70 protons?

b How do the extra neutrons help to keep the isotope stable?

4 Radon-222 is the radioactive gas responsible for much of the Earth's natural background radiation. What type of radiation does it emit? Explain your answer.

5 Describe the main features of the curve for stable isotopes, where number of neutrons (N) is plotted against number of protons (Z).

Higher Questions

10. Alpha, beta and gamma decay 1

By the end of these two pages you should be able to:

- describe the processes of β^- and β^+ decay
- understand gamma decay as an energy loss due to nuclear rearrangement
- state how the proton (atomic) number and mass (nucleon) number change as a result of α, β^-, β^+ and γ decay.

Atoms with too few or too many neutrons can sometimes exist for a while, but they are unstable (**radioactive**). Their nuclei soon decay by emitting radiation, in the form of alpha, beta or positron particles or gamma rays.

Beta (β^- and β^+) decay

Suppose an atom has too many neutrons to be stable, it cannot just throw one out – the neutrons are stuck together too firmly in the nucleus. What it can do is turn a neutron into a proton and an electron. The electron, having hardly any mass, is known as a beta particle (β^-).

You can write β^- decay as:

$$^{1}_{0}\text{n} \rightarrow {}^{1}_{1}\text{p} + {}^{0}_{-1}\text{e}$$

neutron proton β^-

Note that the masses are in the superscripts and the charges are in the subscripts.

The electron has a mass number of 0 and an atomic number of –1 (one negative charge). β^- decay takes place when the unstable hydrogen isotope tritium changes itself into a stable isotope of helium. The nuclear equation is:

$$^{3}_{1}\text{H} \rightarrow {}^{2}_{3}\text{He} + {}^{0}_{-1}\text{e}$$

1 proton 2 neutrons	2 protons 1 neutron	β^- particle

Notice that the mass numbers on each side add up to the same total (3 = 3 + 0), and so do the charges (1 = 2 + –1). This is always true in any nuclear reaction.

When an atom does not have enough neutrons to be stable, a proton turns into a neutron and the nucleus emits a positron (β^+). A positron is the anti-particle equivalent of an electron, with a positive charge:

$$^{1}_{1}\text{p} \rightarrow {}^{1}_{0}\text{n} + {}^{0}_{+1}\text{e}$$

This happens with beryllium-7. It decays to lithium-7. The nuclear equation is:

$$^{7}_{4}\text{Be} \rightarrow {}^{7}_{3}\text{Li} + {}^{0}_{+1}\text{e}$$

4 protons 3 neutrons	3 protons 4 neutrons	β^+ particle

1 Here are four radioactive isotopes of carbon: $^{10}_{6}C$, $^{11}_{6}C$, $^{14}_{6}C$, $^{15}_{6}C$. Two emit electrons and two emit positrons. Which two emit positrons? Explain your answer.

2 Complete the nuclear equation for the positron emission of oxygen-15.

$$^{15}_{8}O \rightarrow {}^{\_}_{\_}N + {}^{\_}_{\_}\_\_$$

Not all radioactive isotopes decay by giving off electrons or positrons. Some release gamma waves.

Gamma (γ) decay

After beta decay, a nucleus is often left in an excited state – that is, with some extra energy. As the protons and neutrons rearrange themselves, they release this energy as an electromagnetic wave, known as a gamma ray.

In gamma decay, the nucleus does not change into a new element because a gamma ray has no mass and no charge.

3 What happens to the mass number in gamma decay? Explain your answer.

Alpha (α) decay

Heavy isotopes, with proton numbers above 82, usually decay by ejecting alpha particles. Alpha particles are the same as helium-4 nuclei, having two neutrons and two protons glued together. Here is a typical example:

$$^{238}_{92}U \rightarrow {}^{234}_{90}Th + {}^{4}_{2}He$$

92 protons 146 neutrons uranium nucleus	90 protons 144 neutrons thorium nucleus	2 protons 2 neutrons alpha particle

4 In the nuclear decay equations on these two pages, what do the superscript numbers refer to?

5 Explain how radioactive isotopes decay into different elements, referring to alpha and beta decays.

A Marie Curie. Marie and Pierre Curie discovered the radioactive elements radium and polonium in some uranium ore.

Summary Exercise

Higher Questions

11. Alpha, beta and gamma decay 2

By the end of these two pages you should be able to:

- state how the proton (atomic) number and nucleon (mass) number change for α, β^-, β^+ and γ decays.

High-energy radiation can damage DNA. This could cause inflammation, cell death or damage to genes, which can lead to mutations in offspring and cancer. Cells which divide frequently (e.g. in the gut walls) are more sensitive than those which rarely divide (e.g. nervous tissue).

The risks associated with radioactive isotopes do not only come from the radiation they emit. When a nucleus emits an alpha or a beta particle, the atom changes into a new element. This daughter element may be hazardous. For example dangerous radon gas comes from the disintegration of radium and is a major part of natural background radiation. It is dangerous to breathe in radon gas because it emits alpha particles, which can damage DNA and cause lung cancer.

Since alpha particles only travel a short distance in living tissue they give up their energy quickly. This makes the relative danger of alpha particles greater than that of beta particles or gamma rays.

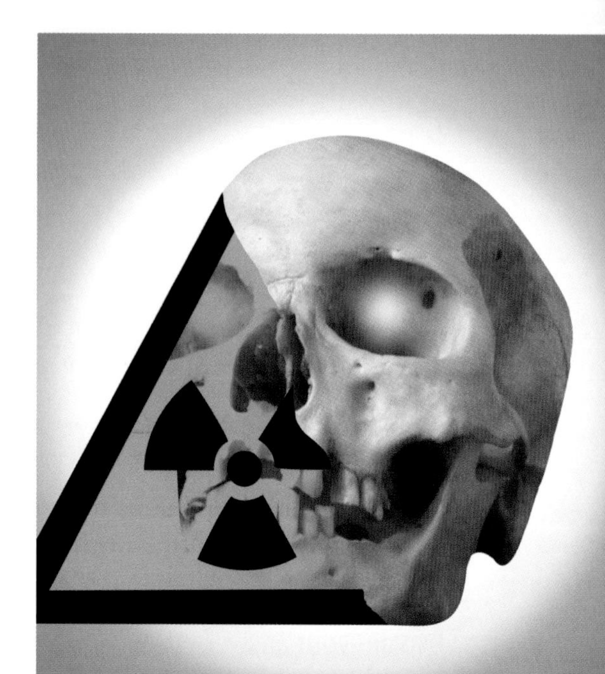

A Improper use of radioactive materials can be fatal.

1 a Which type of radiation causes more damage by bombarding cells in our bodies?
b Why could radon damage your lungs?
c How many protons and how many neutrons are there in radium-226 ($^{226}_{88}Ra$)?
d What is the atomic number and mass number of the radon isotope produced when radium-226 decays? (Remember an alpha particle contains two protons and two neutrons.)

2 a What is beta-minus radiation?
b What happens to the mass (nucleon) number and the proton (atomic) number during beta minus decay?
c When an atom emits a positron, how does its nucleus change?

After decay, the changes in the nucleus result in a movement on the graph of neutron number (*N*) against proton number (*Z*).

If a decay step causes the nucleus to return to the belt of stability on the *N–Z* plot, the daughter element is stable and not radioactive.

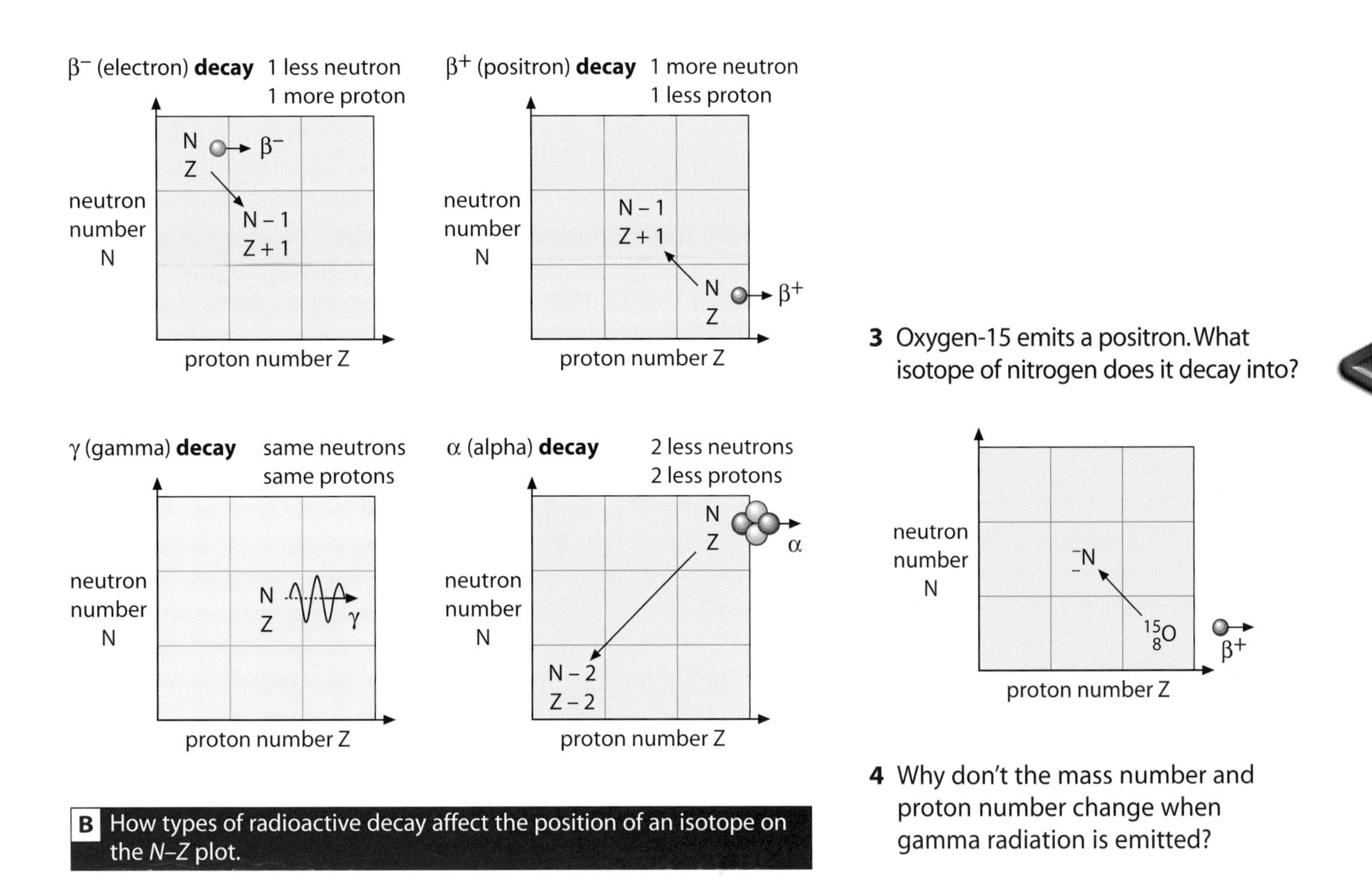

B How types of radioactive decay affect the position of an isotope on the *N–Z* plot.

3 Oxygen-15 emits a positron. What isotope of nitrogen does it decay into?

4 Why don't the mass number and proton number change when gamma radiation is emitted?

5 Copy and complete the table summarising the differences between a, β^-, β^+ and γ decays.

Radiation	Symbol	Problem with nucleus	Change as a result of decay		
			Nucleus	**Mass number**	**Proton number**
alpha	$\alpha = {}^{\_}_{\_}\text{He}$	More than ___ protons (helium nucleus emitted)	2 protons, 2 ___ leave	down	down 2
beta-minus (electron)	$\beta^- = {}^{\_}_{-1}\text{e}$	Too ___ neutrons (____ the *N–Z* curve)	___ changes to proton (and emits an electron)	no change	___ 1
beta-plus (_______)	$\beta^+ = {}^{0}_{\_}\text{e}$	Too few neutrons (____ the *N–Z* curve)	___ changes to neutron (and emits a positron)	___	down ___
Gamma	γ	Too excited	No change, but emits an electromagnetic wave	no change	________

Summary Exercise   Higher Questions   Extension Questions

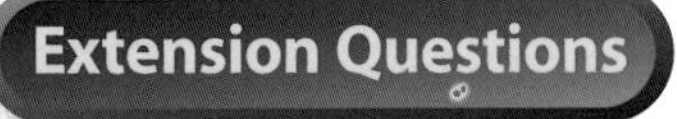

12. Nuclear decay equations

By the end of these two pages you should be able to:

- write nuclear equations for α, β^-, β^+ and γ decay.

When studying topic P2.12 Power of the atom, you probably looked at a diagram showing **nuclear fission**.

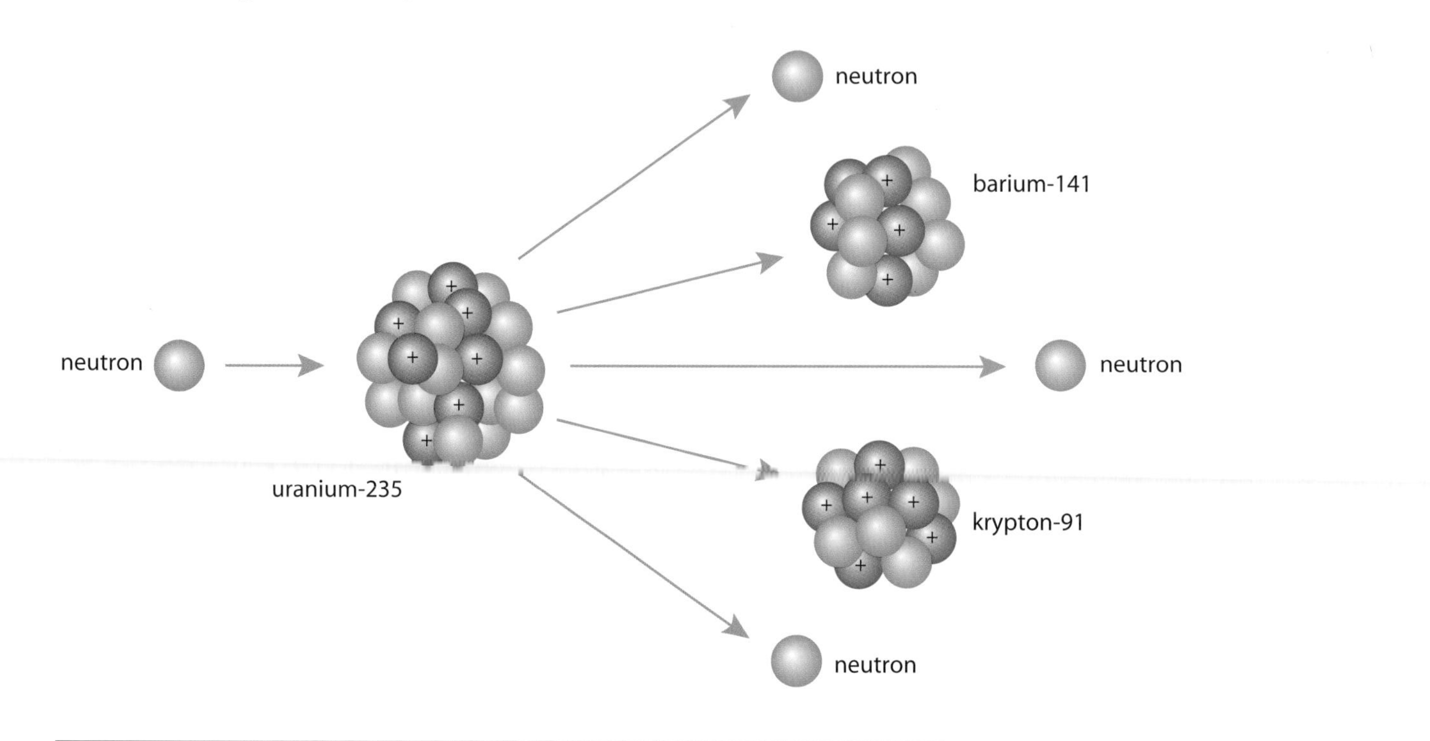

B An example of the fission of uranium-235. Other fission products are possible.

A nuclear equation shows the same data:

(143 + 92 = 235)

$$^{1}_{0}n + ^{235}_{92}U \rightarrow ^{236}_{92}U \rightarrow ^{143}_{56}Ba + ^{90}_{36}Kr + 3\,^{1}_{0}n$$

(92 protons)

(143 neurons)

Add mass numbers: 1 + 235 = 236 = 143 + 90 + 3

Add atomic numbers: 0 + 92 = 92 = 56 + 36 + (3×0)

Note the mass numbers and the atomic (or charge) numbers on each side add up to the same total.

Unlike a chemistry equation, in a nuclear equation you are only interested in showing the change to a nucleus, rather than changes to molecules. One danger of nuclear power is the hazardous radioactive fission fragments. Both barium-143 and krypton-90 emit beta minus particles (electrons). Krypton-90 also emits gamma radiation. Nuclear radiation can damage DNA.

1 a How many neutrons are in a krypton-90 nucleus?

b Is krypton-90 above or below the *N–Z* curve of graph B in lesson P3.5 9 on page 171?

c Does krypton-90 emit β^- particles because its nucleus has too many neutrons or too few?

d Why does a krypton-90 nucleus also emit gamma radiation?

β⁻ decay: When a barium-143 emits an electron the nuclear equation is:

$$^{143}_{56}\text{Ba} \rightarrow {}^{143}_{57}\text{La} + {}^{0}_{1}\text{e}$$

(56 protons) (57 protons) β⁻
(87 neutrons) (86 neutrons)

An electron has hardly any mass compared to a proton or neutron. The element with 57 protons is lanthanum, symbol La. Remember that a neutron becomes a proton in β⁻ decay.

You write the gamma decay equation for krypton-90 as:

$$^{90}_{36}\text{Kr} \rightarrow {}^{90}_{36}\text{Kr} + \gamma$$

Notice there is no change to the nucleus.

β⁺ decay: Look at the nuclear equation for the positron emission of oxygen-15.

$$^{15}_{8}\text{O} \rightarrow {}^{15}_{7}\text{N} + {}^{0}_{+1}\text{e}$$

(8 protons) (7 protons) β⁺
(7 neutrons) (8 neutrons)

A positron has the same mass as an electron, which is not much at all! The oxygen nucleus changes to a nitrogen nucleus. Notice that a proton becomes a neutron in β⁺ decay.

α decay: In alpha decay a nucleus emits two protons and two neutrons, glued together like a helium nucleus.
Smoke alarms contain americium-241. Smoke stops the alpha particles reaching the detector. The decay equation is:

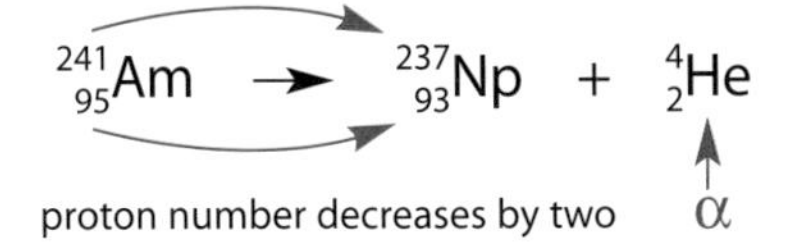

The element with 93 protons is neptunium, symbol Np.

4 Write the nuclear equation for the alpha decay of the gas radon-220 ($^{226}_{86}$Rn) into polonium (Po).

In 1986 the world's worst nuclear power accident occurred at Chernobyl in the Ukraine. Whilst testing a reactor, numerous safety procedures were disregarded. Explosions blew off the reactor's heavy steel and concrete lid, killing 30 people immediately and over 2500 later on. In regions contaminated by radioactive waste, nine times as many children develop thyroid cancer and three times as many children are born with deformity.

2 Write the nuclear equation for krypton-90 (Kr), when β⁻ decay changes it into rubidium (Rb).

3 Complete the nuclear equation for fluorine-18, a positron emitter.

$$^{\_}_{9}\text{F} \rightarrow \text{O} + {}^{\_}_{\_}$$

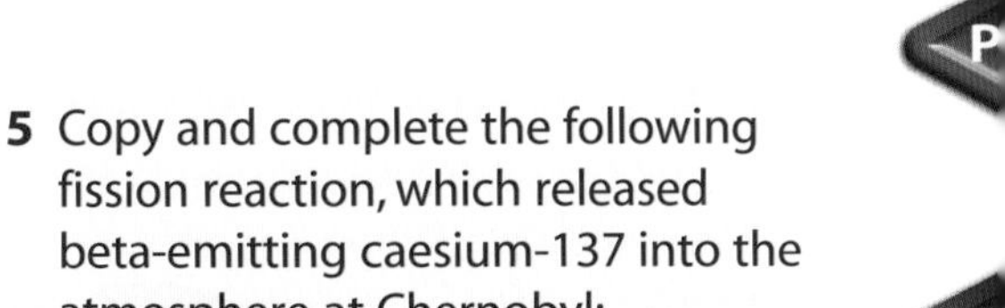

5 Copy and complete the following fission reaction, which released beta-emitting caesium-137 into the atmosphere at Chernobyl:

$$^{0}_{1}\text{n} + {}^{235}_{92}\text{U} \rightarrow {}_{56}\text{Cs} + {}^{95}\text{Rb} + {}^{1}_{0}\text{n}$$

6 Explain how to write a nuclear equation.

Summary Exercise | Higher Questions | Extension Questions

13. Quarks

By the end of these two pages you should be able to:

- account for the number of up and down quarks in protons and in neutrons in terms of charge and mass
- describe how instruments, such as particle accelerators, can help us to explain the world.

In 1964, American physicist Murray Gell-Mann suggested a strange idea. He said that protons and neutrons are each made from three quarks (pronounced 'quorks') glued together, and quarks have one of two 'flavours' – *up* or *down*. Five years later he won the Nobel prize. Experiments at CERN, the European centre for particle research, and elsewhere have now confirmed his idea. The oddest thing is that, compared to an electron with a negative charge $-e$, an up quark has a charge $+{}^{2}/_{3}e$ and a down quark has a charge $-{}^{1}/_{3}e$ So quarks have fractions of the charge of an electron. Not only do quarks exist but so do their anti-particles, called **anti-quarks**. These have opposite properties to their equivalent quark.

1 Choose the correct flavour of quark: up or down.
 a Which quark has a negative charge?
 b Which quark has the bigger charge?

How could three quarks combine to make a proton or a neutron? The possibilities are shown in diagram C.

2 **a** Which particle has a combination of quarks uud (up up down)?
 b Which particle has a combination of charges: $+{}^{2}/_{3}, -{}^{1}/_{3}, -{}^{1}/_{3}$ Explain your answer.
 c Which type of quark is there more of in a neutron?

3 How many quarks are there in an alpha particle ($^{4}_{2}$He)?

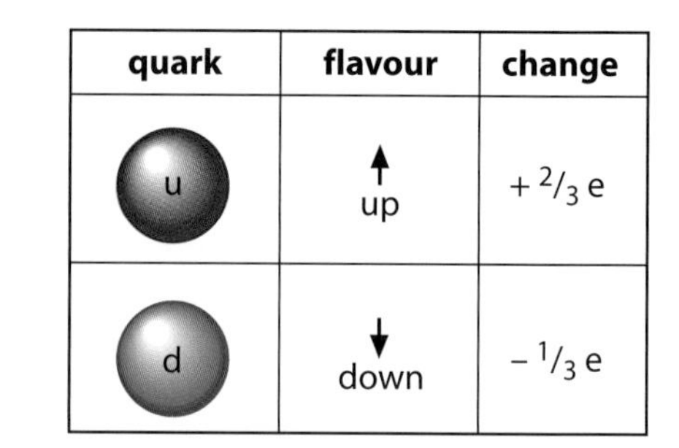

quark	flavour	change
u	up	$+{}^{2}/_{3}$ e
d	down	$-{}^{1}/_{3}$ e

A Charge and flavour of quarks.

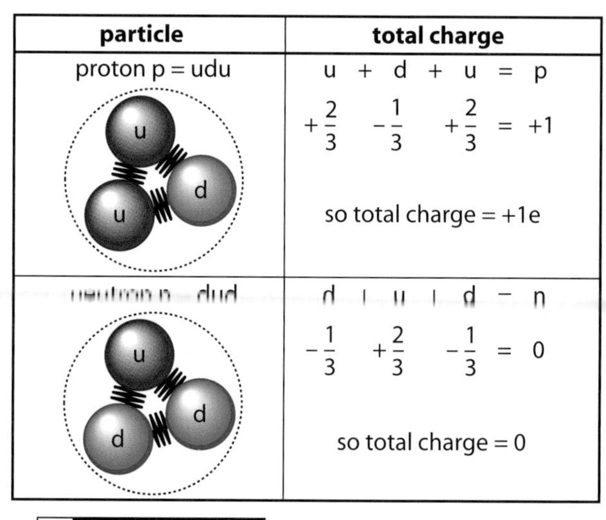

particle	total charge
proton p = udu	u + d + u = p $+\frac{2}{3} \quad -\frac{1}{3} \quad +\frac{2}{3} = +1$ so total charge = +1e
neutron n = dud	d + u + d = n $-\frac{1}{3} \quad +\frac{2}{3} \quad -\frac{1}{3} = 0$ so total charge = 0

B Quark summary.

C Ducks say 'Quark, Quark'! Quarks explain protons and neutrons.

How do physicists study particles?

Particles are tiny, and to 'see' them, scientists need accelerators. These huge machines speed up particles to very high kinetic energies, before smashing them into other particles and breaking them apart. Scientists then carry out experiments to study the effects of the collisions.

Particle accelerators are either linear (straight-line) or circular.

- In linear accelerators (or linacs) the particle is shot like a bullet from a gun. Linacs range in size from the simple electron guns in X-ray machines and oscilloscopes to the 3 km long Stanford linear accelerator in California.
- In circular accelerators (or synchrotrons) the particle track is bent into a circle by electromagnets. For example, the Large Hadron Collider particle accelerator at CERN.

Diagram D shows electrons moving through a linear particle accelerator. The linac works by switching the voltage of the tubes from positive to negative so the particles accelerate forward at successive gaps.

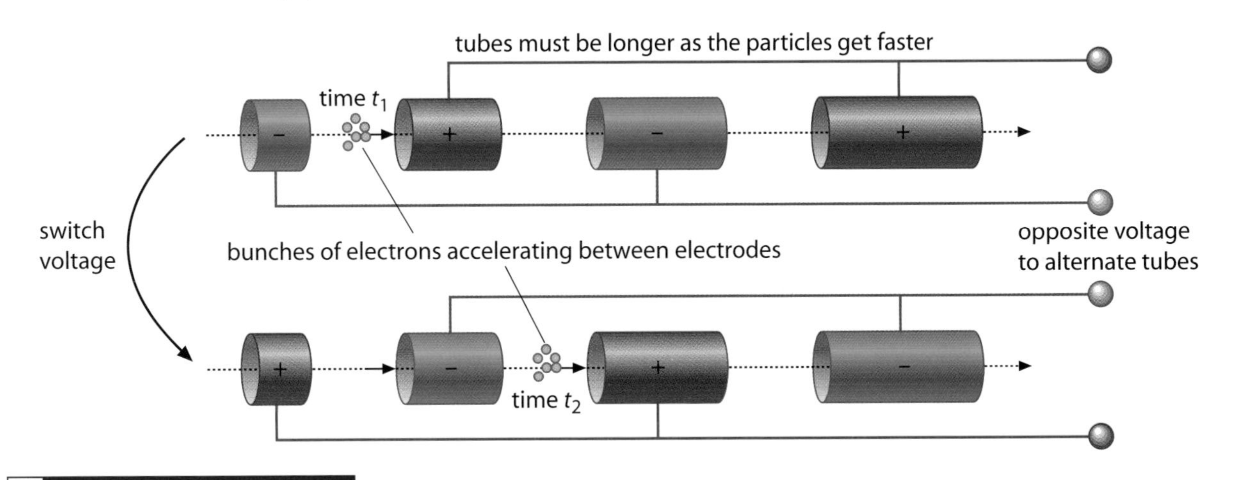

D A linear particle accelerator.

To accelerate positive particles like protons, the tube behind the protons would need to be positive to repel them forwards, with those in front negative to attract them. Scientists can confirm the existence of quarks by smashing protons together at near the speed of light and detecting the particles produced.

4 In a linear particle accelerator:
  a why do electrons accelerate towards a positive tube
  b how does the time the particles spend between each tube change?

5 Murray Gell-Mann had no direct evidence when he suggested the existence of quarks. What happens in a particle accelerator to provide that experimental evidence?

Summary Exercise

Higher Questions

14. Fundamental particles

By the end of these two pages you should be able to:

- state what is meant by, and give examples of, fundamental particles, such as electrons and quarks, and their anti-particles
- explain that protons and neutrons are not fundamental particles because they each contain three quarks
- discuss how scientists create anti-matter such as a positron
- describe the properties of a positron.

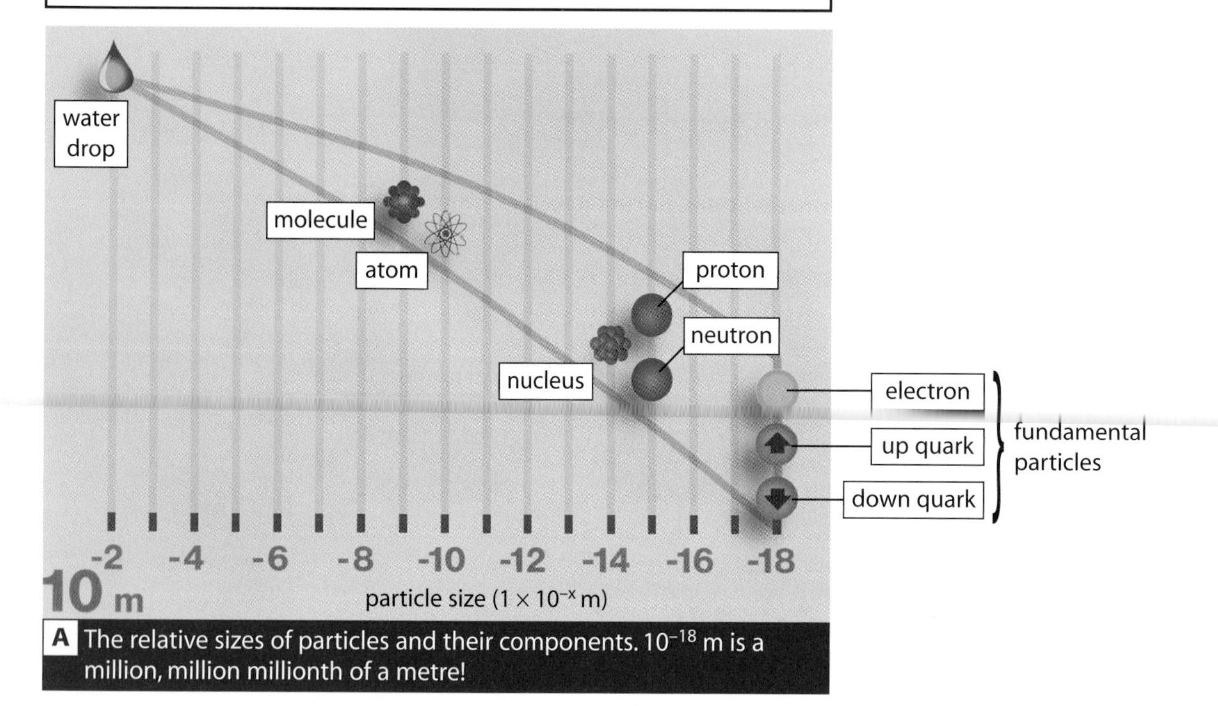

A The relative sizes of particles and their components. 10^{-18} m is a million, million millionth of a metre!

You know that a drop of water is composed of water molecules. A water molecule (H_2O) is made from hydrogen and oxygen atoms. An oxygen atom contains protons, neutrons and electrons. Protons and neutrons are each made from three quarks. A **fundamental particle** is one which cannot be broken down any further. Scientists believe that quarks and electrons are fundamental particles.

The anti-matter equivalents of quarks and electrons are anti-quarks and positrons. These are also fundamental particles, and like all particles are held together by nuclear or electrical forces.

An anti-particle has the same mass as its corresponding particle but an opposite charge. Electrons are negative, positrons are positive, but they have the same mass. As a result they travel in opposite directions in electrical or magnetic fields. You already know that if a particle and its anti-particle come into contact, like an electron and a positron, they destroy each other. The annihilation causes two gamma rays to rush away in opposite directions at the speed of light.

1 Why are protons and neutrons not fundamental particles?

2 Name six fundamental particles.

Doctors use electron–positron annihilation in **PET (positron emission tomography)** scans of the brain (see P3.6 Medical physics). Before the scan, you inject a patient with glucose containing the positron emitter fluorine-18. The more active parts of the brain absorb more of this glucose. Electrons in surrounding atoms immediately annihilate any positrons emitted, allowing gamma cameras to pinpoint where the bursts of activity take place.

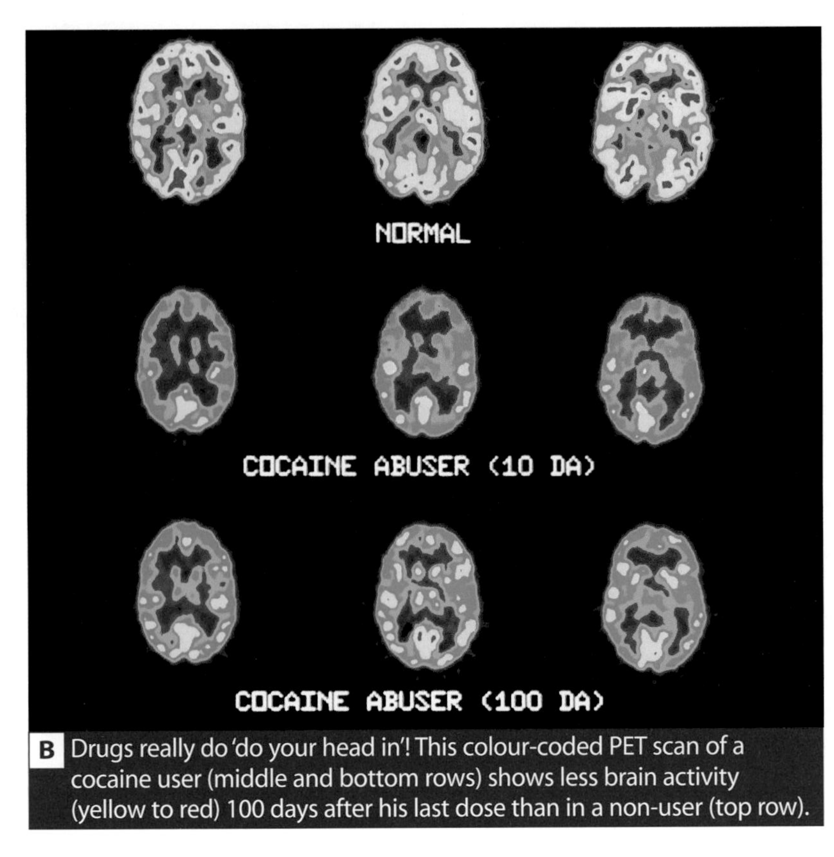

B Drugs really do 'do your head in'! This colour-coded PET scan of a cocaine user (middle and bottom rows) shows less brain activity (yellow to red) 100 days after his last dose than in a non-user (top row).

Fluorine-18 does not occur naturally. However, you can manufacture $^{18}_{9}F$ by bombarding water molecules containing an oxygen isotope $^{18}_{8}O$ with protons, which you first accelerate to high speeds in a machine called a cyclotron. The reaction is given by the equation:

$$^{18}_{8}O + ^{1}_{1}H \rightarrow ^{18}_{9}F + ^{1}_{0}n$$

proton neutron

Have you ever wondered?

Is anti-matter real or just science fiction?

Have you ever wondered?

Is there anything smaller than protons, neutrons and electrons?

3 99.8% of naturally occurring oxygen is $^{18}_{9}O$ with eight protons and eight neutrons. How many neutrons are there in the $^{18}_{8}O$ isotope?

4 Explain how the symbol $^{1}_{1}H$ means just one proton.

5 A by-product in the manufacture of fluorine-18 is neutron radiation. Neutrons are difficult to stop. Explain whether or not this radiation poses a hazard. (*Hint*: see lesson P3.5 8.)

6 Explain, using examples, why some particles are called fundamental particles and some are not.

15. Quarks changing flavour!

By the end of these two pages you should be able to:

- describe β^- decay as a down quark changing into an up quark (one neutron becomes a proton and an electron)
- describe β^+ decay as an up quark changing into a down quark (a proton becomes a neutron and a positron)
- describe how instruments, such as particle accelerators, can help us to explain the world.

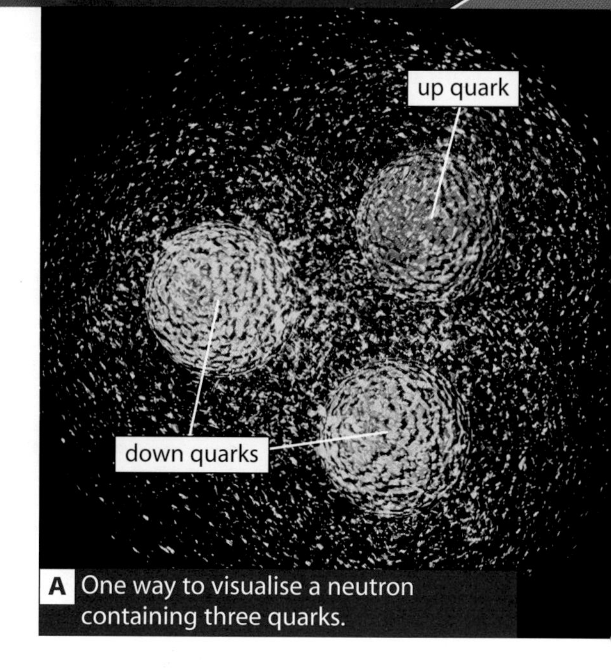

A One way to visualise a neutron containing three quarks.

You may be thinking, 'Beta particles are electrons. Electrons orbit the nucleus. The nucleus contains protons and neutrons, not electrons. How can a beta particle be emitted from the nucleus in a radioactive decay?' Remember Murray Gell-Mann's strange idea about protons and neutrons containing three quarks. Amazingly, in radioactive isotopes a quark can change its flavour.

In isotopes with too many neutrons, a down quark becomes an up quark. This causes a neutron to change into a proton. This change of flavour from down to up creates an electron, which is emitted as a beta particle.

In isotopes with too few neutrons, an up quark becomes a down quark. This causes a proton to change to a neutron. This change of flavour from up to down creates a positron.

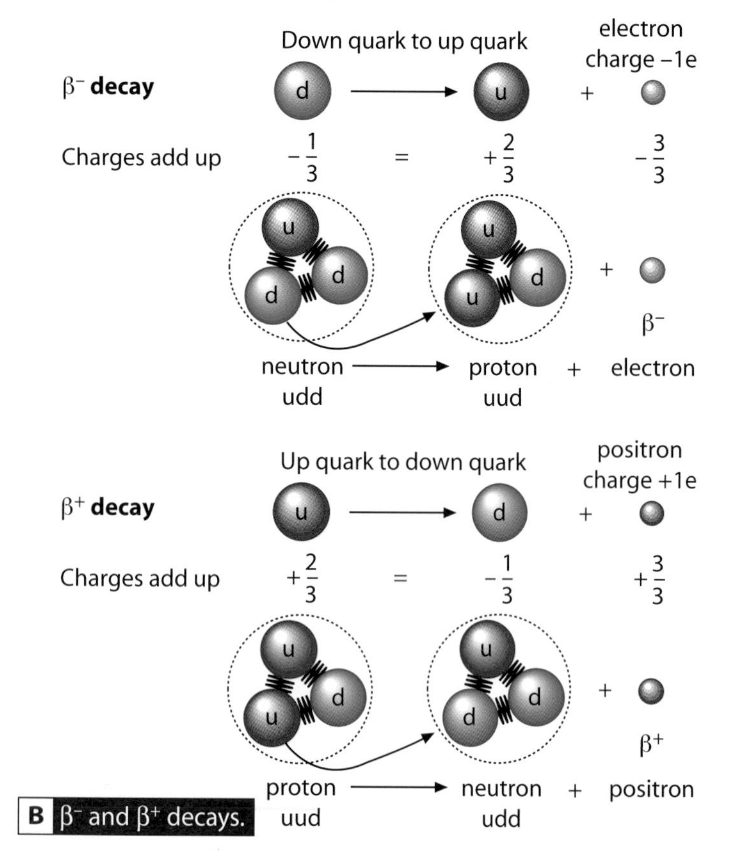

B β^- and β^+ decays.

1 An up quark changes its flavour and becomes a down quark. What particle does the nucleus emit?

2 A nucleus has too many neutrons. What change of flavour can remedy this? Explain your answer.

3 An up quark has charge $+^2/_3$. A down quark has charge $-^1/_3$. What is the difference between their charges?

4 Electrons normally orbit the nucleus of an atom. Explain how an electron can suddenly leave the nucleus of a radioactive atom.

During beta decay, quarks change flavour. This makes the isotopes more stable. In lesson P2.11 6 you measured the half-life of a radioactive sample – the average time taken for half the atoms to decay. This time differs from one isotope to the next, and not all the nuclei in a given isotope decay at the same time. Nuclear decay is a random process that follows a definite pattern. For example, in positron decay a certain up quark may change to a down quark in the next second, while another up quark in the same isotope may wait a lifetime before it changes flavour. Nevertheless within a time, equal to the half-life, you can expect half the quarks in the isotope to change flavour.

5 Electrons normally orbit the nucleus of an atom. Explain how an electron can suddenly be fired out of the nucleus of a radioactive atom.

What questions remain?

To the surprise of physicists, particle accelerator experiments have revealed many new discoveries.

- There are six types of quark, with the odd names of up, down, strange, charm, bottom and top. Quarks never exist separately, but only in pairs or in threes (like in protons and neutrons). Such quark combinations are called hadrons (or heavy particles).
- There are six types of leptons (or light particles) like the electron and electron-neutrino.
- For every particle there is an equivalent anti-matter particle.

Yet many questions remain unanswered.

- Is there any pattern to the different masses of the quarks and leptons?
- Are the quarks and leptons really fundamental particles, or do they, too, have sub-structure?
- Are there more types of particles to be discovered with accelerators?

Perhaps higher-energy collisions can provide clues to these questions.

C The quark family.

16. Synchrotrons and CERN

By the end of these two pages you should be able to:

- explain the reasons for collaborative, international research into big scientific questions, such as particle physics.

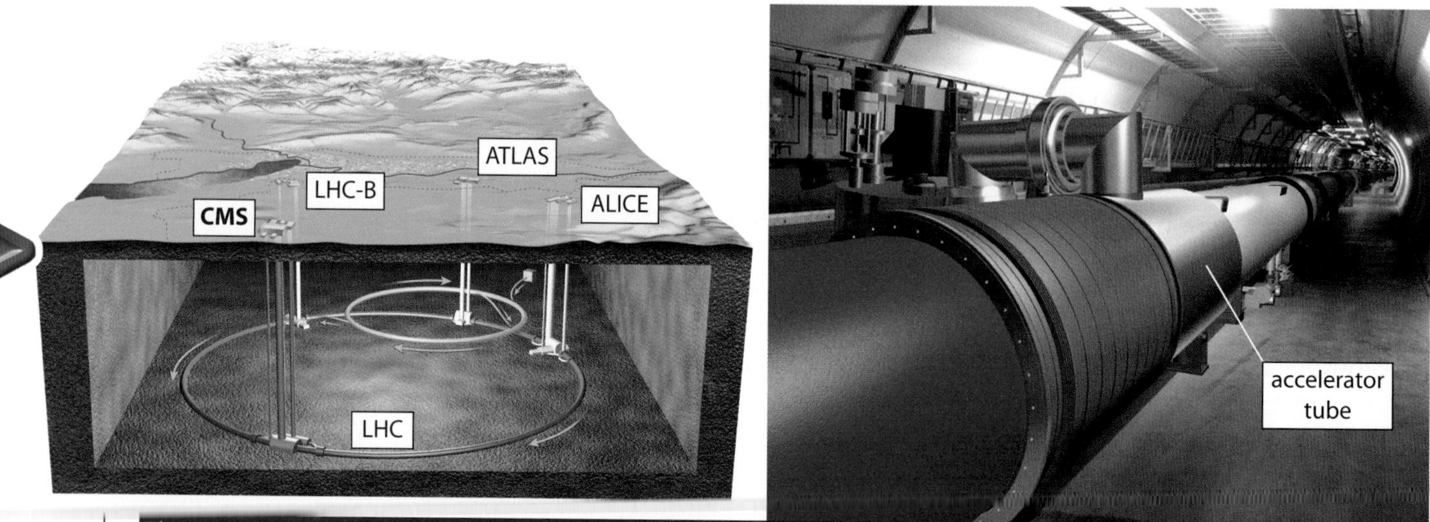

A The Large Hadron Collider (LHC) at CERN. The four collision sites to detect particles are ATLAS, ALICE, CMS and LHC-B.

Quarks and electrons are tiny, smaller than 10^{-18} m across – that is one million, million, millionth of a metre. Are more types of particles, even smaller, yet to be discovered? The biggest experiment in history, CERN's **Large Hadron Collider (LHC)**, begins operating in 2007. It is likely to shed light on this and some of the most fundamental questions in science – from the formation of stars to everything you see around you. The LHC is a 27 km long circular particle accelerator (or synchrotron) buried 100 m underground and crossing the French–Swiss border near Geneva. It has two pipes, down which groups of protons or heavy ions accelerate in opposite directions, approaching the speed of light. As the particles collide, huge detectors record the nature of the particles produced. The LHC is probing deeper into matter than ever before, with energies that only existed at the birth of the universe. It represents a US$7 billion investment that has cost the UK taxpayer £34 million per year over the past ten years. Thousands of people and half the world's particle physicists, from 80 nations, are collaborating to analyse the results.

Technological spin-offs from work already carried out at CERN include the invention of the World Wide Web and the production of particle beams for radiation therapy in hospitals. Tim Berners-Lee invented the World Wide Web in 1990 to help CERN particle physicists communicate via computers on a global scale. Today the global community relies on it – and the rest, as they say, is Web history.

1 Quark combinations like protons are called hadrons.
  a What is the Large Hadron Collider?
  b What are scientists recording when the hadrons collide?
  c What answers do scientists hope to find out from the LHC experiments?

The trick to creating a circular beam is to bend the charged particles with electromagnets. However, if the magnetic field remains constant while you accelerate the particles, the beam spirals out. This is much like forcing a marble to speed up as it spins round a bowl. A synchrotron overcomes the spiralling effect and keeps the particles in a circle by:

- increasing the magnetic field strength, and
- matching the frequency of the accelerating voltage pulses to the rate of rotation of the particles.

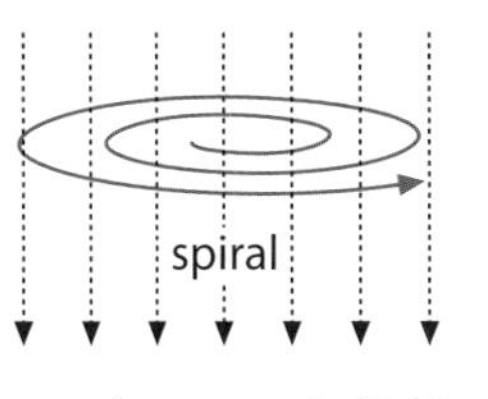

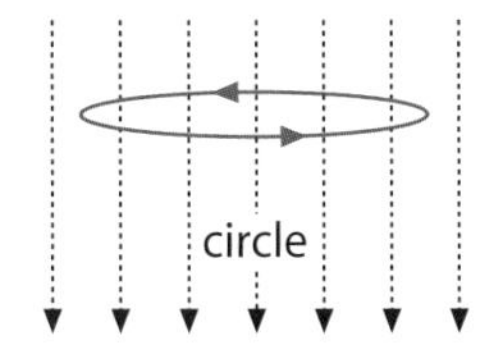

B Accelerating particles in a magnetic field.

2 What factors limit the speed to which you can accelerate particles in a synchrotron?

At CERN, huge kinetic energies are possible because the LHC has:

- a large circumference (27 km)
- powerful, superconducting electromagnets
- very high alternating frequencies to accelerate the particles to almost the speed of light
- two beams accelerating in opposite directions before the particles collide.

3 What is the purpose of the super-conducting electromagnets in the LDC?

4 How do you change the voltage to accelerate a bunch of protons though the accelerating cavity?

5 How is collaborative, international research taking place at CERN?

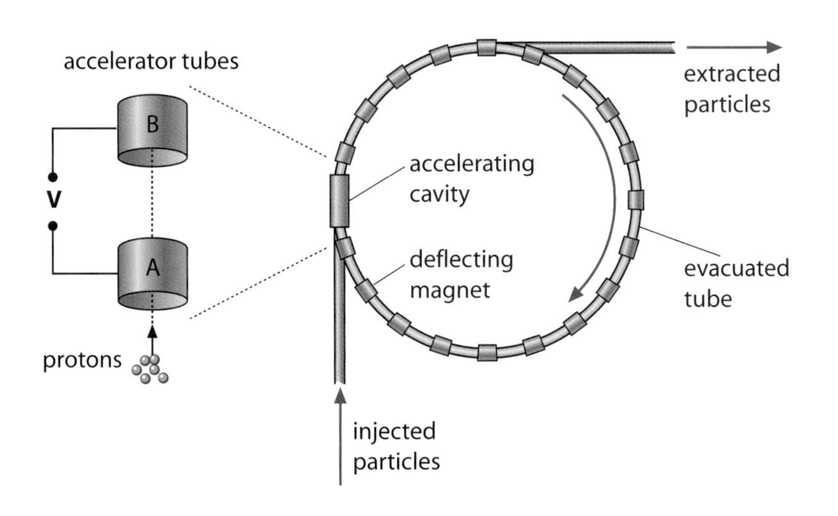

C The LHC at CERN is a synchrotron in the shape of a doughnut.

The accelerating cavity of the LHC is simply two charged cylinders (A and B in diagram D). The kinetic energy of the protons is increased by changing the voltage when the protons enter A and leave B.

Have you ever wondered?

Why do some scientists spend their lives on an experiment consisting of 27 km of empty space?

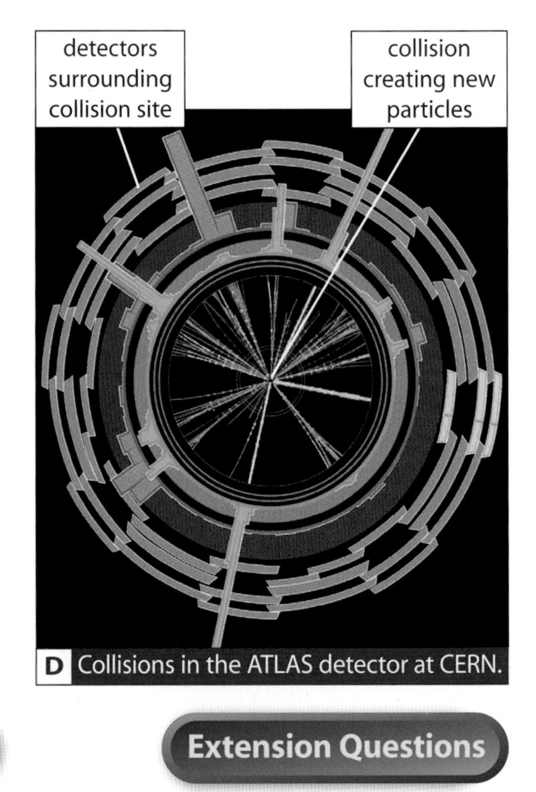

D Collisions in the ATLAS detector at CERN.

Extension Questions

17. Questions

1 a Gases are affected by temperature and pressure. A car tyre contains air at a pressure of 1.2×10^5 Pa when the temperature is 27 °C.

i Explain how the air exerts a pressure on the inside walls of the tyre.

ii Convert 27 °C to a temperature on the kelvin scale.

b Once the car has been running for a while the temperature of the air in the tyre rises to 82 °C.

i State how this affects the average kinetic energy of the particles of air.

ii If the volume of the tyre does not change, what is the new pressure of the air in the tyre?

iii If the tyre expands by 5% as its temperature rises, how does the new pressure compare with your answer to part **b ii**?

c If the car tyre is taken and dumped in outer space, explain whether or not it would explode.

2 a Beams of electrons may be produced by an electron gun and carry energy that may be converted into X-rays.

i Use the following words to name parts A–D on the diagram of an X-ray machine:

heated cathode   accelerating anode
electron   X-ray

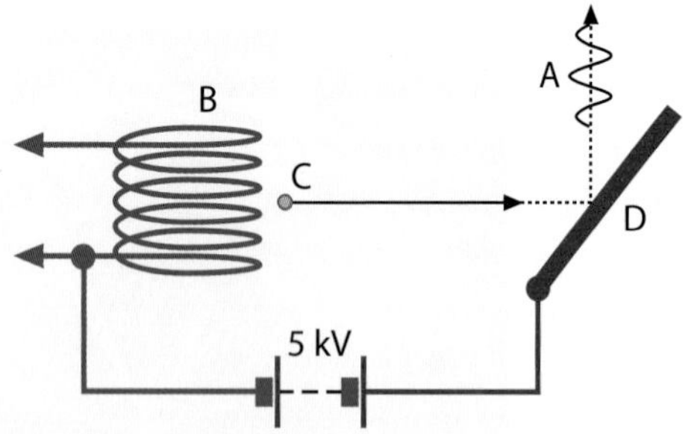

ii Explain the process that causes electrons to be 'boiled off' the cathode.

iii What property of the anode attracts the electron?

iv Why must the electron gun be mounted in a vacuum?

b The accelerating voltage of the electron gun is 5 kV. Given the charge on an electron is 1.6×10^{-19} C, with what kinetic energy do the electrons arrive at the anode?

c If 6×10^{17} electrons 'boil off' the cathode each second, what is the current of the electron beam?

d State three ways in which a television tube differs from an X-ray tube.

3 Unstable isotopes and their emissions may be identified by the position of the isotope on a neutron/proton curve.

a A is the unstable isotope indium-116, with 49 protons.

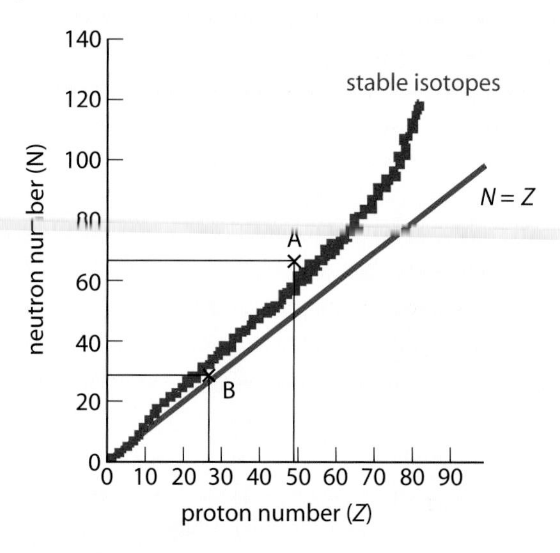

i What is an isotope?

ii What is the neutron number of indium-116?

iii What does it mean that indium-116 is 'unstable'?

iv Indium-116 is a β^- emitter. What is a β^- particle?

v Complete the equation showing the decay of indium-116:

$^{116}_{49}\text{In} \rightarrow \text{Sn} + {}_{-1}$—

b B is an isotope of cobalt with 29 neutrons that emits a positron when it decays.

i What is a positron?

ii Complete the equation showing the decay of this isotope:

$^{\_}_{27}\text{Co} \rightarrow {}^{\_}_{\_}\text{Fe} + {}^{\_}_{\_}$

iii What does a proton within the cobalt nucleus become as a result of the decay?

iv Explain what happens to the quarks of this proton when the positron is emitted.

4 Electron beams are used in a variety of equipment including oscilloscopes.

a Explain how an electron beam deflects in an oscilloscope.

b What do you use to bend the particles in a circular particle accelerator (or synchrotron)?

c A group of positive ions enter the accelerating cavity of a synchrotron. The voltage then changes to accelerate the ions forward from A to B.

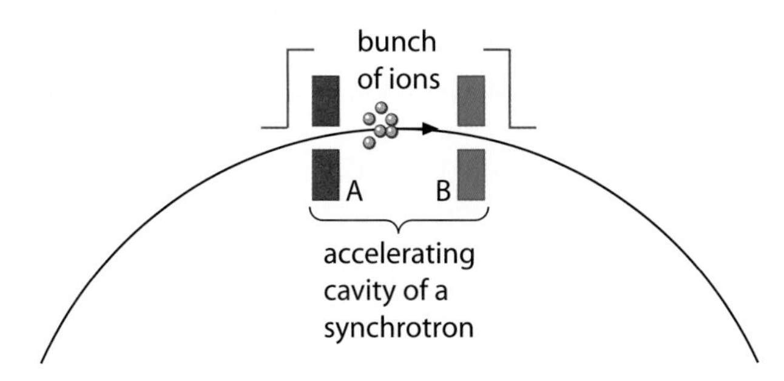

Consider an ion at the front of the group.

i Is the time it takes to reach electrode B more or less than the time taken by an ion at the back?

ii Is its increase in speed on reaching B more or less than that of an ion at the back?

iii Why do ions get more tightly bunched during acceleration?

d The Large Hadron Collider (LHC) at CERN is a synchrotron.

i How are scientists using the LHC to study fundamental particles?

ii Give three examples of fundamental particles.

18. Glossary

D

***absolute zero** The lowest temperature possible when all the energy has been taken away from the particles. It is –273 °C, or 0 K (kelvin).

***accelerating anode** A positive electrode that increases the speed of negatively charged particles. It is used in an electron gun.

***alpha particle** A particle made of two protons and two neutrons. It is emitted from heavier radioactive isotopes with proton numbers greater than 82.

anti-matter Composed of anti-particles.

anti-particle Has the opposite charge but same mass as its corresponding particle, which it annihilates on contact.

anti-quark The anti-particle equivalent of a quark.

belt of stability The line made up of the stable isotopes on an *N–Z* graph (a graph of number of neutrons against number of protons) for all the elements.

***beta particle** High kinetic energy electron (β^-) or positron (β^+), emitted from radioactive isotopes with too many or too few neutrons.

***cathode** A negative electrode.

cathode ray tube (CRT) An evacuated glass tube containing an electron gun. The electrons ('cathode rays') originate from the hot cathode.

***Celsius** The temperature scale with units °C.

***decay** The process of transforming to another element or isotope when a radioactive isotope emits radiation.

electron A negative particle that normally orbits the nucleus of an atom. However, a β^- particle is an electron emitted from a nucleus in radioactive decay when a neutron changes into a proton.

***electron beam** A stream of negatively charged particles, used in oscilloscopes, TV tubes and X-rays.

evacuated Where the air has been removed so there is a vacuum.

***fundamental particle** A particle that cannot be broken down into smaller units. At present quarks, electrons and their anti-particles (anti-quarks and positrons) are thought to be fundamental.

***gamma radiation** Pulses of electromagnetic waves with very short wavelengths, emitted from radioactive nuclei that are in an excited state.

***isotopes** Atoms of the same element having a different number of neutrons in their nucleus – e.g. carbon-12 (six protons, six neutrons); carbon-14 (six protons, eight neutrons).

***kelvin** The temperature scale starting with 0 K at absolute zero. K = 273 + °C.

***kinetic energy** The energy an object has because of its speed.

Large Hadron Collider (LHC) The 27 km long synchrotron (circular particle accelerator) at CERN used for colliding hadrons (particles made up of quarks).

***neutron** The neutral particle in the nucleus of atoms, made up of one up quark and two down quarks.

nuclear fission The splitting of a nucleus by neutrons. It occurs when uranium-235 fragments in a nuclear power station.

nucleon Particles such as protons and neutrons that make up the nucleus.

***nucleus** The centre of an atom containing protons and neutrons, around which electrons orbit.

***oscilloscope** An instrument that displays how voltages are changing with time.

***particle** The general name for a minute portion of matter, ranging in size from electrons and quarks to atoms and molecules.

***particle accelerator** A machine used to produce a high-energy particle bunch or beam. Accelerators are either linear (straight-line) called linacs, or circular called synchrotrons.

***positron** The anti-particle of an electron, having the same mass but opposite charge. Positrons are emitted from isotopes with too few neutrons.

positron emission tomography (PET) A medical scanning technique. The image of metabolically active sites inside the body is computed by detecting gamma rays coming from positron–electron annihilation.

***pressure** A force acting over a certain area, that can be caused by the collisions of the particles of a gas on a surface.

***proton** The positive particle in the nucleus of atoms, made up of two up quarks and one down quark.

***quark** Particle from which protons and neutrons are made. Protons and neutrons are hadrons (heavy particles) containing three up and down quarks.

***radiation** The emission of energy from a source, as sound, electromagnetic waves or moving particles.

***radioactive** When the unstable nucleus of an atom decays.

radioactive isotope Unstable isotope that emits radiation, as alpha particles, beta particles (electrons and positrons), gamma rays or neutrons.

***temperature** The degree of hotness of a body, measured on the Celsius or kelvin scale. (K = 273 + °C.)

***thermionic emission** The process of emitting (or 'boiling off') an electron when a hot atom changes to a positive ion.

time base The control on an oscilloscope that controls the speed the electron beam moves across the screen.

*glossary words from the specification

Medical physics

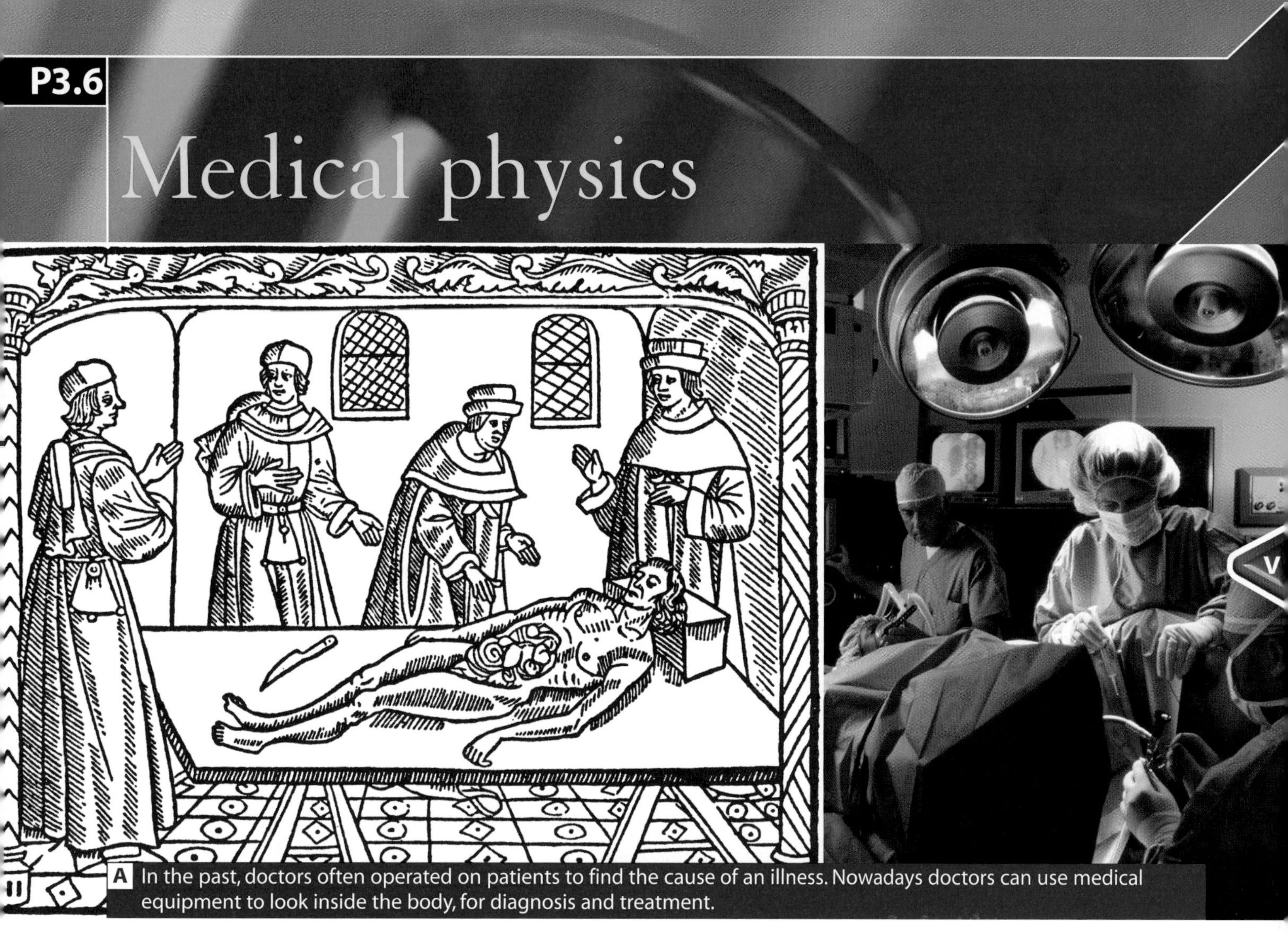

A In the past, doctors often operated on patients to find the cause of an illness. Nowadays doctors can use medical equipment to look inside the body, for diagnosis and treatment.

Medical physics is the use of physics in medicine. Knowledge of physics concepts has led to the development of sophisticated medical equipment that can be used to examine inside the body, often saving the need to carry out surgery.

In this topic you will learn about applications of physics to medicine, so you will cover areas of both physics and biology. Radiation is used in the diagnosis and treatment of cancer. The reflection and refraction of light are used to look inside the body and measure pulse rate and blood oxygen levels. The electrical activity of the heart can be monitored to see whether a heart is healthy or not. Scanners that can be used to diagnose the body make use of particle physics. When new techniques are tested, there are ethical issues involved.

In this topic you will learn that:

- structures and organs inside the body may be examined without cutting a patient open
- radiation affects living matter and can be used to destroy malignant tumours
- new medical techniques can raise moral and ethical issues.

Sort the following statements into three categories:

true, not true, not sure.

- Radiotherapy kills normal body cells as well as cancer cells.
- When light moves from air to water it changes speed, called reflection.
- A pulse oximeter detects electricity in your heart to measure pulse rate.
- ECGs are only recorded for patients at risk of a heart attack.
- When a particle and an antiparticle collide this is called annihilation.

1. Human energy

By the end of these two pages you should be able to:

- recall that work done is equal to energy transferred
- recall and use the equation:
  work done = force × distance moved in the direction of the force, $W = F \times s$.

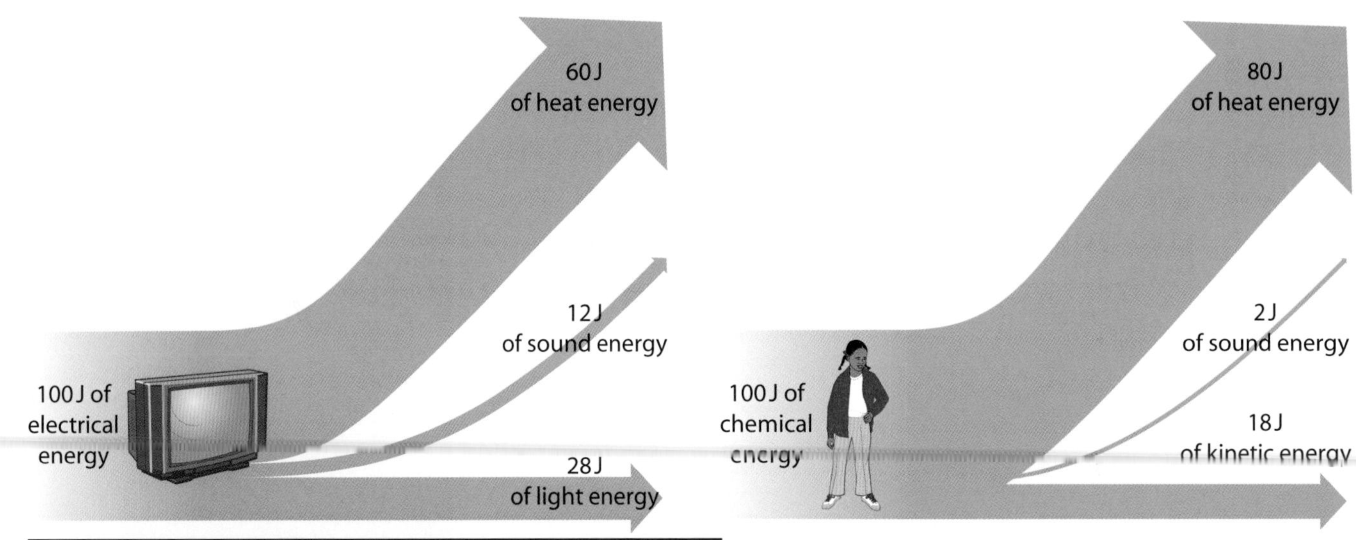

A Televisions and children transfer energy into different types.

In order to function, a television takes in electrical energy. It transfers this energy to light energy and sound energy. Most televisions also give off quite a lot of heat energy. All of the electrical energy entering will be transferred to other types of energy.

The human body is also a machine. We take in chemical energy in food and drink. We use this so we can move (kinetic energy) and make a noise, such as talking (sound energy). Like the television, we give off heat to our surroundings. In fact each person produces about the same amount of heat as a 100 W light bulb.

Person	Typical energy required (J per day)
newborn baby	2 000 000
2-year-old child	6 000 000
teenage girl	10 000 000
teenage boy	12 000 000
office worker	11 000 000
heavy construction worker.	15 000 000
pregnant woman	10 000 000
breast-feeding woman	11 000 000

B How much energy do I need each day?

1 What is 15 kJ, expressed in J?

2 Look at diagram A.
   a In what ways are the energy transfers in a television and a child similar?
   b Where do the 100 J of chemical energy supplied to the child come from?
   c Which of the three forms of energy transferred by the television are useful?

3 Look at table B.
   a How much energy do you need each day?
   b Why do you think a teenage boy needs more energy than a teenage girl?

All types of energy are measured in joules (J). It is quite a small unit, so kilojoules (kJ) are commonly used. 1 kJ is equal to 1000 J. One joule of **work** is the energy required to move one metre against a force of one newton. Therefore one joule is the same as one newton metre (Nm). Work is done when there is movement against any type of friction, against the force of gravity, or when a charged particle is repelled as it moves in an electric field. The energy is transferred to other types of energy, such as heat and sound.

You can calculate work done using the following equation: work done, W (J) = force, F (N) × distance moved in the direction of the force, s (m)

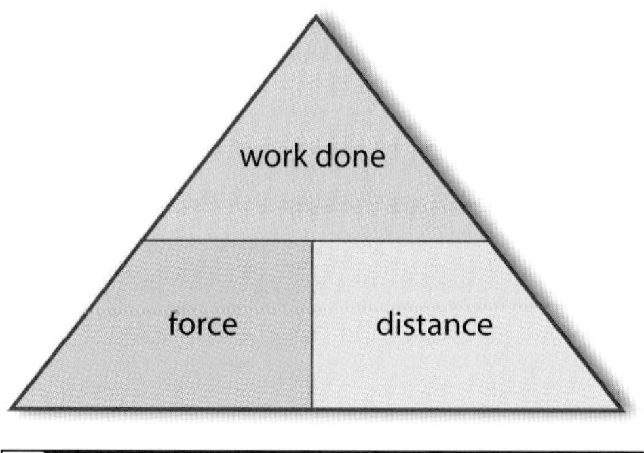

C Work, force and distance memory triangle.

Work done against friction

In diagram D, the girl pulls the suitcase for 40 metres. She needs to overcome the force of friction, so her average force is the same as the force of friction: 120 newtons.

work, W = force, F × distance moved in the direction of the force, s
= 120 N × 40 m
= 4800 Nm = 4800 J

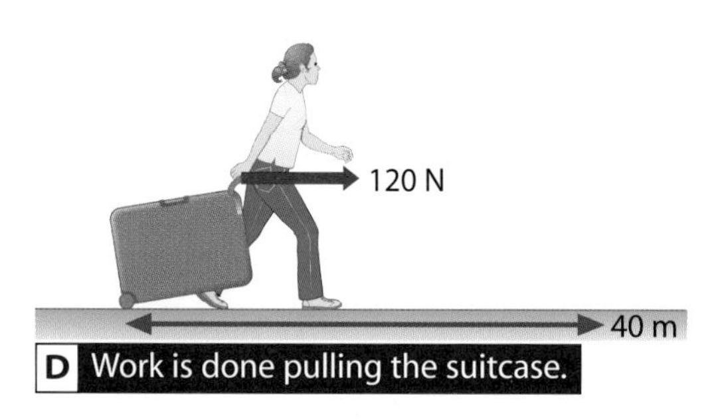

D Work is done pulling the suitcase.

Work done against weight

In diagram E, the boy lifts the paint from one shelf to the other through a height of 0.3 metres. The weight of the tin is 20 newtons.

work, W = force, F × distance moved in the direction of the force, s
= 20 N × 0.3 m
= 6 Nm = 6 J

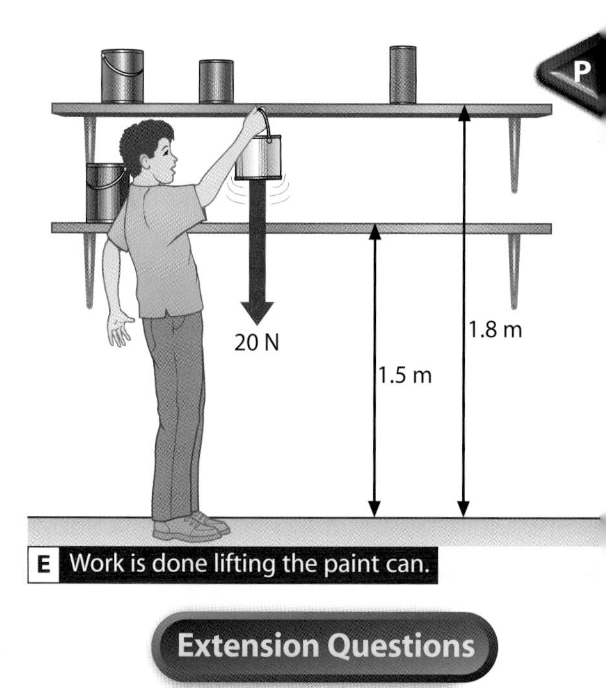

E Work is done lifting the paint can.

4 What is 30 J, expressed in Nm?

5 How much energy is transferred when a car moves 80 m against a friction force of 600 N?

6 How much energy is transferred when a bag of 5 N is lifted through a height of 40 cm?

7 Energy cannot be created or destroyed, it is transferred from one type of energy to another. Look at diagrams D and E. Draw energy diagrams showing how 100 J of chemical energy might be transferred in each case.

Summary Exercise
Higher Questions
Extension Questions

2. Human power

By the end of these two pages you should be able to:

- recall and use the equation power = work done/time taken, $P = W/t$
- understand the idea of basal metabolic rate (BMR).

How much electrical energy is used by a light bulb? That will depend on the length of time it is on for, and on the **power** of bulb. The word power is used in different ways in everyday life. To scientists, power is the rate of transfer of energy and is measured in watts. 1 kilowatt (kW) is equal to 1000 watts (W). A kettle is described as having a power of 2200 W. This means that it uses 2200 J of electrical energy each second. Therefore one watt is the same as another unit, the joule per second.

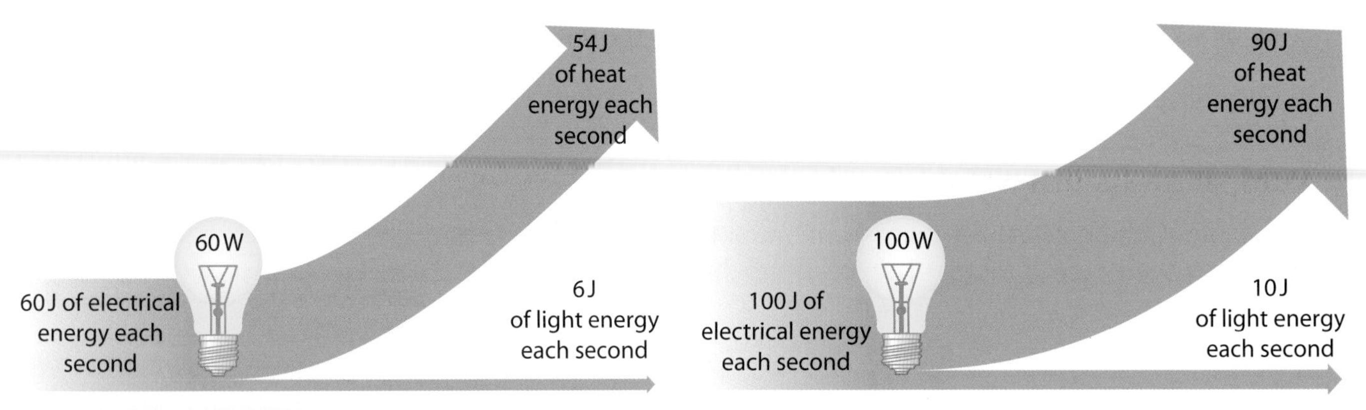

A Which bulb is brighter: 60 W or 100 W?

Power can be calculated as energy transferred over time, and since energy transferred is the same as work done, power is also equal to work done divided by the time taken.

$$\text{power}, P\ (\text{W}) = \frac{\text{energy transferred}, E\ (\text{J})}{\text{time taken}, t\ (\text{s})}$$

$$\text{or power}, P\ (\text{W}) = \frac{\text{work done}, W\ (\text{Nm})}{\text{time taken}, t\ (\text{s})}$$

Remember: $W = F \times s$

1 What is 8500 W expressed in J/s?

2 What is 8500 W expressed in kW?

3 An electric shower uses 840 000 J of electrical energy in 2 minutes. What is its power? (*Hint*: change the minutes to seconds).

4 What is the power of a motor which lifts a 800 N object through a height of 4 m in 20 s?

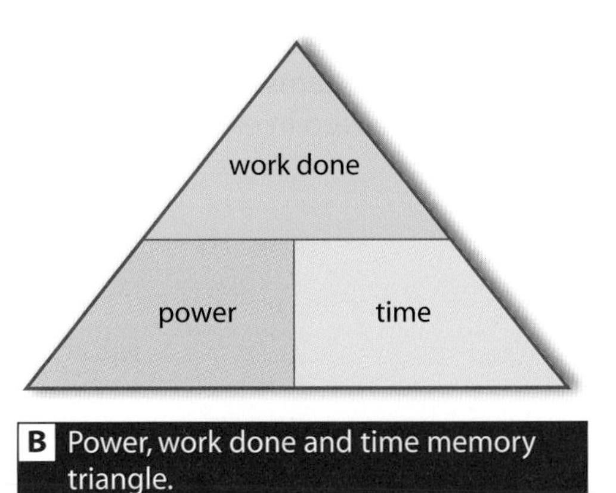

B Power, work done and time memory triangle.

C The amount of energy transferred each second depends on the activity.

Activity	Typical power (W)
sitting	90
standing	120
walking slowly	210
getting dressed	240
walking quickly	350
playing tennis	430
playing football	600
running	700

D Energy conversion rates for different activities.

Even when lying in bed, it is impossible to use no energy. In this inactive state you need energy to make the movements involved with breathing, to make your heart contract about once a second and to carry out the many chemical reactions, such as making enzymes, which are needed to keep you alive. Energy is needed to keep these basic processes functioning.

The minimum amount of energy on which the body can survive is called the **basal metabolic rate (BMR)**. The BMR is about 7 million joules each day. Slightly more energy is needed if you are young, because your body is making new cells at a greater rate than in the body of an adult, and BMR also varies with the sex and health of an individual.

A young person at rest has a BMR of 7 000 000 joules per day. What is his power (in J/s)?

power = energy transferred to other types of energy/time taken

$$= \frac{7\,000\,000 \text{ J}}{24 \times 60 \times 60 \text{ s}}$$

$$= 81 \text{ J/s}$$

5 A footballer needs 1.6 million joules for a 45 minute training session. What is his power?

6 Using the figures for typical powers in table D, work out how long you should do each activity in order to convert 100 000 J.

Higher Questions

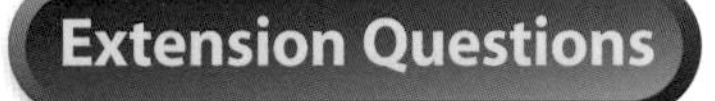

3. Radiating energy

By the end of these two pages you should be able to:

- describe and explain the term intensity
- know and be able to use the equation intensity = power of incident radiation/area, $I = P/A$
- use the word 'radiation' to describe any form of energy originating from a source
- know that the intensity of a radiation will decrease with distance from a source
- know that the intensity of a radiation will decrease through a medium.

In science, **radiation** is used to mean energy spreading out from a **source**. Radiation can be the harmful emissions from **radioactive** materials, but can also be everyday radiation such as light from a bulb – the light energy radiates outwards from the source.

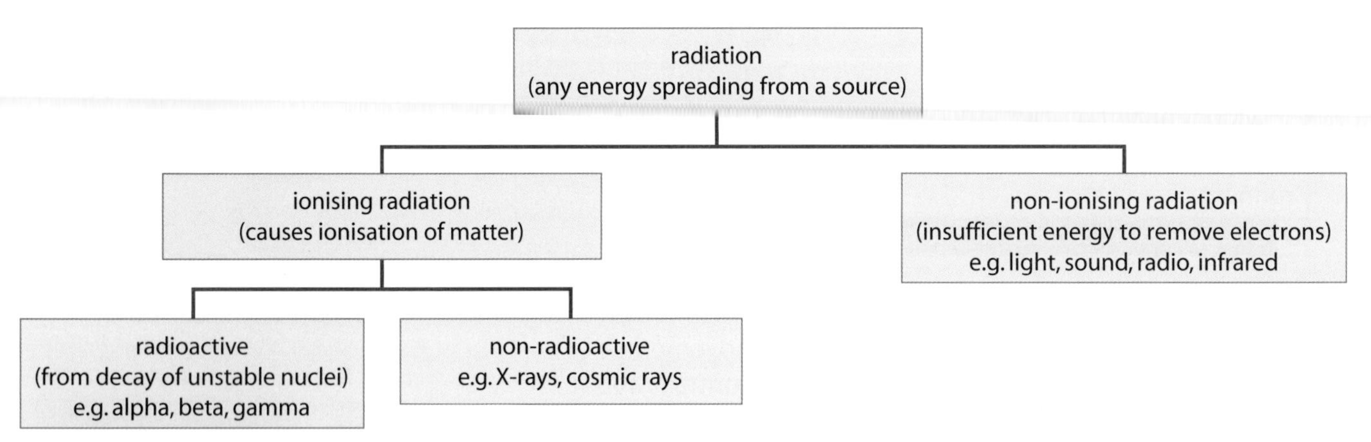

A Types of radiation.

Ionising radiations cause **ionisation** of matter. They have sufficient energy to knock electrons out of the atoms of a material, forming **ions**. Radiations such as light and sound do not have sufficient energy to remove electrons, and so are called non-ionising.

1 Why is a stream of alpha particles described as ionising?

2 Gamma rays, X-rays and light are electromagnetic waves. How are they similar and how do they differ?

3 Do all ionising radiations come from radioactive decay?

The strength of a radiation is called its **intensity**. This is the power of incident (striking a surface) radiation per square metre and is measured in W/m^2 or mW/m^2 (1 W = 1000 mW).

$$\text{intensity}, I = \frac{\text{power}, P}{\text{area}, A}$$

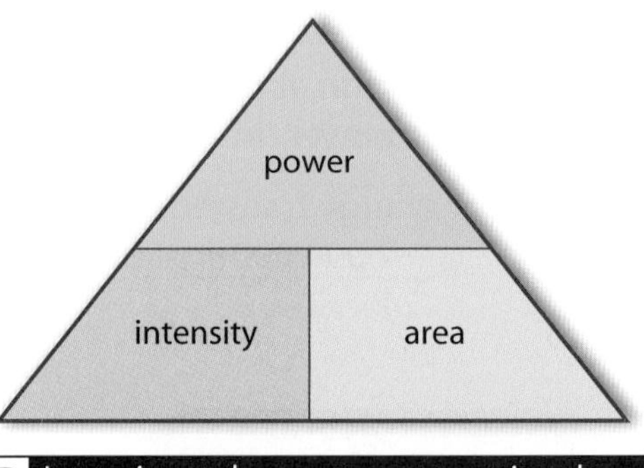

B Intensity and power memory triangle.

Energy spreads out, getting less intense, as it moves away from a source, as shown in diagram C. In the orange sphere the radiation has spread out to an area of one small square. The green sphere is twice as far from the source as the orange sphere. The energy has spread out further, to an area of four squares. The red sphere is three times as far from the source as the orange sphere. The radiation has spread out to cover nine squares.

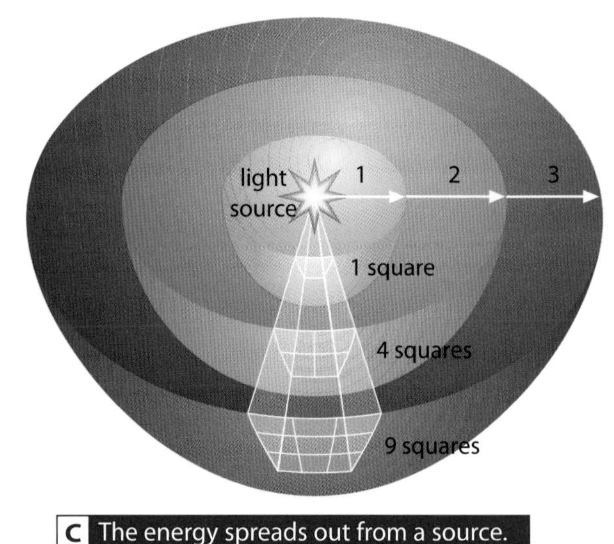

C The energy spreads out from a source.

If the source is emitting radiation continuously, the amount of energy passing through the orange sphere per second is the same as the amount of energy passing through the green and red spheres per second.

If the intensity at the first square (on the orange sphere) is 72 mW/m^2, then all of that energy will continue into the green sphere, over an area of four squares. The 72 mW will be shared equally between four squares, i.e. each square will have 18 mW. This means that the green sphere radiation is a quarter of the original intensity. When the radiation reaches the red sphere, the original energy is now shared between nine squares, that is 8 mW per square. The intensity has decreased to a ninth of its strength at the orange sphere.

Distance from source	1 radius	2 radii	3 radii
Intensity (mW/m^2)	72	18	8
Total area of beam	1 square	4 squares	9 squares

D How intensity of radiation decreases with distance.

When the radiations used for medical purposes pass through the body, they are not in a vacuum, they are in a medium. However, there is a similar pattern of decreasing intensity. Each material in the body reduces the intensity of the radiation by different amounts. Bone is dense and reduces the intensity the most. Muscle is less dense so it does not decrease the intensity as much.

4 Using the numbers in the table above, find mathematical relationships between distance, intensity and area.

5 The graph shows an experiment investigating how intensity depends upon thickness of material. Use the graph to find the thickness of material needed to halve the intensity.

6 Make a concept map to summarise how intensity of radiation varies with distance.

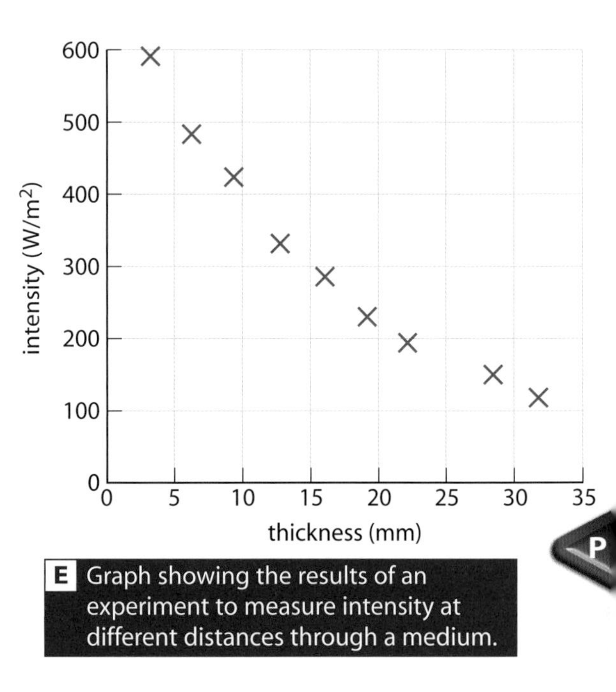

E Graph showing the results of an experiment to measure intensity at different distances through a medium.

Summary Exercise

Higher Questions

Extension Questions

4. Harmful radiation

By the end of these two pages you should be able to:

- describe and explain the effects of radiation on living matter
- describe the importance of limiting exposure to radiation.

When we absorb radiation, the amount received is the **absorbed dose**, measured in joules per kilogram. Ionising radiations differ greatly in the way in which they affect biological materials. For example, alpha radiation causes much more harm than beta radiation for the same amount of energy. The unit which incorporates both the energy and the biological harm is the **sievert** (Sv), or the millisievert (1 Sv = 1000 mSv). It is known as the equivalent dose.

Type of exposure	Equivalent dose of radiation (mSv)
flight from London to Madrid	0.01
chest X-ray	0.02
average annual dose from natural radiation in UK	3
annual dose in Cornwall from natural radiation	8
maximum permitted annual dose for an employee	20
dose which causes immediate radiation sickness	1000
dose causing death in 50% of population	4000

A Absorbed doses from different sources.

Alpha, beta and gamma radiations travel at high speeds, and alpha and beta are electrically charged. When these radiations enter your body they can harm living cells. Low doses ionise the molecules which make up our cells. Thankfully, alpha radiation, which could cause the greatest amount of ionisation, is large and not very penetrating. It is stopped by the outer layer of skin. If the dose is very high, so much energy is transferred that it can burn.

Beta and gamma radiations are able to penetrate deep into the body. Even at low levels, the radiations can ionise DNA. This may cause the cells to grow and divide in an uncontrolled manner and become cancerous. If the ionisation occurs in the DNA of the sperm or egg, then the result could be children born with birth defects. However, the body has systems which check the quality of DNA, and gametes with defective DNA are usually destroyed.

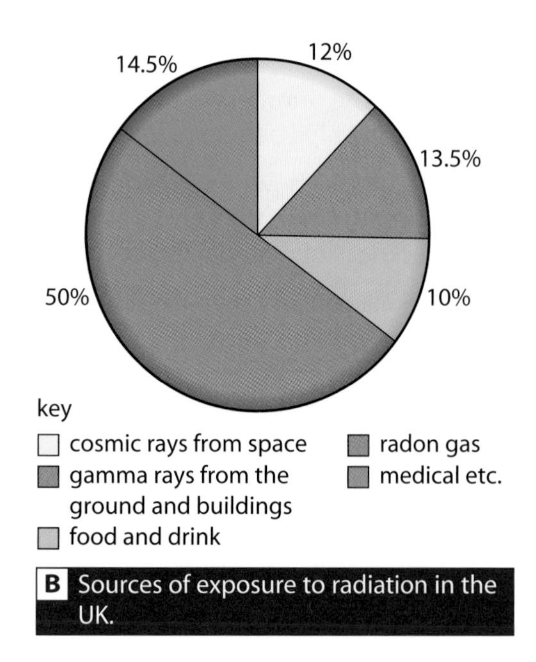

B Sources of exposure to radiation in the UK.

1. How do ionising radiations affect the human body?
2. What are the three main sources of radiation?
3. What percentage of the radiations passing through a typical person in the UK are natural and cannot be avoided?

On average, about half of all radiation comes from radon, a radioactive gas found particularly in areas with large amounts of granite and shale in the ground. Radon is the second largest cause of lung cancer after smoking. However, diagram C shows that you are more likely to die from an accident indoors than from lung cancer caused by radon.

Cosmic rays originate from outer space and collide with molecules in the atmosphere. At altitude there are more cosmic rays because fewer of them have collided with air molecules, so the exposure of air crew and frequent fliers can be 100–300 times higher than the exposure at sea level. The average person in the UK receives 2.7 mSv of ionising radiation a year, nuclear workers receive 3.6 mSv and air crew receive 4.6 mSv. These levels are still much lower than the European annual limit of 20 mSv set for a worker in a nuclear power plant. Research has shown that male pilots have slightly higher rates of several cancers compared with the national average. However, women seem to be at greater risk. Air hostesses have twice the risk of breast cancer compared to the average flier. Nevertheless, statistics show that you are more likely to be healthy and safe in a plane than doing many other activities. So, unless you are pregnant, you should not fly less to avoid cosmic rays.

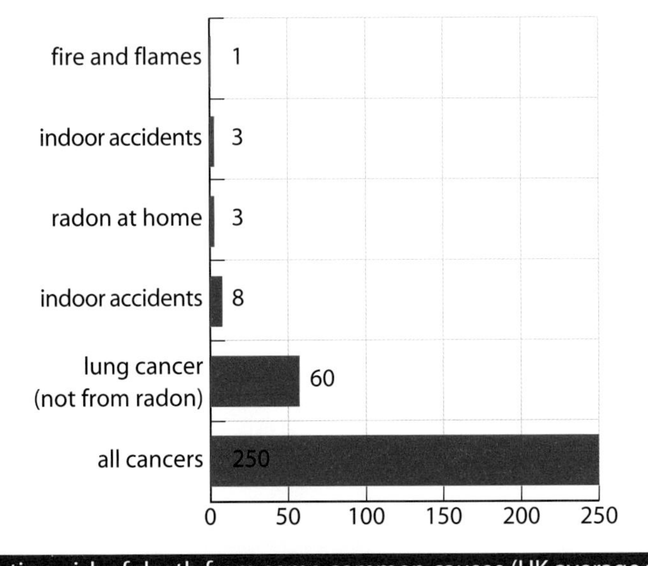

C Lifetime risk of death from some common causes (UK average for smokers and non-smokers per 1000 people) (source: www.hpa.org.uk).

4 What can you do to reduce your exposure to radiation?

5 Why should women who are frequent fliers consider having regular health checks?

6 Write a leaflet for residents living in a high-radon area telling them about radiation, the risks from radon and how it is important to reduce their exposure.

Higher Questions

Extension Questions

5. Radiotherapy

By the end of these two pages you should be able to:

- explain the application of radiation in treating malignant tumours
- explain that radiation treatment does not always lead to a cure and that it is sometimes used to reduce suffering (palliative care).

Have you ever wondered?

How do we use radiation to treat cancer?

There are many different types of cancer, all caused by abnormal cell growth. Normally cell division is carefully controlled, but with cancer the cells start dividing in an uncontrolled way. This can result in a lump of rapidly dividing cells called a **malignant tumour**. Some malignant tumours spread around the body, circulating in the blood or lymph vessels. **Benign tumours** are not cancerous. They stay in one place and do not harm surrounding tissues, except by pressing on them.

Radiotherapy can be used to treat or to control cancer. Radiotherapy uses **gamma rays**, particles such as electrons or, more commonly, a beam of high-energy X-rays. Electron beams do not penetrate well, so can be used to treat skin cancer. X-rays have a higher energy so can penetrate tissues more easily and travel further. They can be used to treat a **tumour** inside the body.

The aim of radiotherapy is to destroy cancer cells whilst keeping any damage to healthy cells to a minimum. Carefully targeted, measured doses of radiation can be used for a range of conditions including thyroid disorders and blood diseases, as well as cancers. Radiotherapy is used on its own to treat thyroid cancer and some brain tumours. For other conditions it can be used, together with surgery, to reduce the size of the tumour before it is removed, or after an operation to make sure that all cancerous cells are destroyed. It can also be used with chemotherapy (the use of anti-cancer drugs).

Although radiotherapy can be given on a single occasion, it is usually given in several small doses over a period of days or weeks. Our bodies are made up of cells, which divide to make new cells for growth or repair. If radiation hits a cell that is dividing it might be damaged. Cancer cells are more likely to be dividing and are much less able than normal cells to repair the damage. So cancer cells are damaged much more by radiotherapy than normal cells. Any healthy cells which are affected are quickly replaced by normal cells.

1 What is cancer?

2 How are malignant tumours different from benign tumours?

3 What radiations are used in radiotherapy?

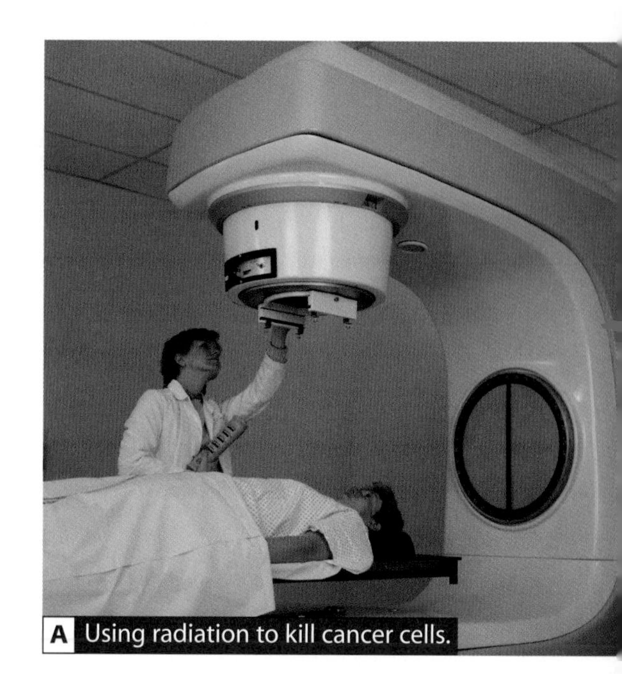

A Using radiation to kill cancer cells.

The beams are focused to ensure that the target area, rather than healthy cells, receives the most radiation. One approach is to direct the rays from different directions. The cancer is hit by every beam, but healthy tissue is only exposed to the one beam that passes through it.

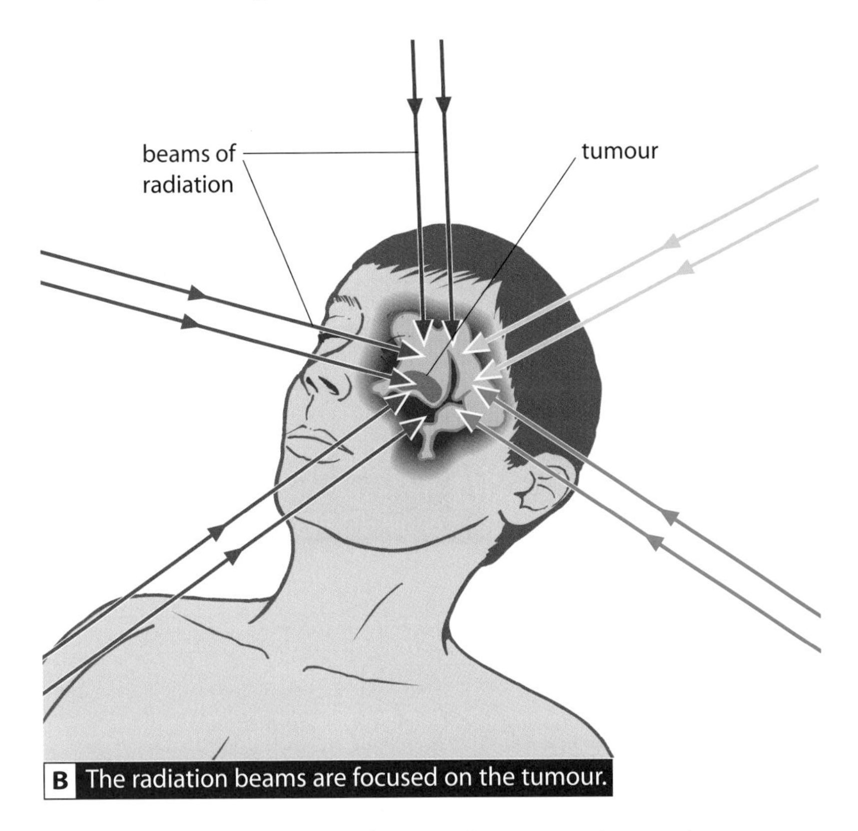

B The radiation beams are focused on the tumour.

Some treatments cannot destroy the cancer. Instead radiotherapy is intended to control the symptoms of cancer. This is known as **palliative care**. The radiotherapy might be used to shrink a tumour which is producing pressure, causing a blockage or causing compression on the spinal cord. This approach is also used on tumours which have grown under the skin and have produced wounds. The radiotherapy can shrink the tumour, stopping any discharge or bleeding.

4 What is the difference between chemotherapy and radiotherapy?

5 What is meant by palliative care?

6 Explain two ways that damage to healthy cells is kept to a minimum during radiotherapy.

7 Draw a poster or annotated diagram to show how radiotherapy is used to treat malignant tumours.

Higher Questions

6. Reflection and refraction

By the end of these two pages you should be able to:

- describe what is meant by refraction.

When light hits a wall or other flat object, it bounces off at the same angle. This is **reflection** and it allows us to see objects which do not produce their own light – we see the reflected rays which enter our eyes. Diagram A shows light striking a flat surface. The angle at which the ray meets the surface (the **angle of incidence**) is the same as the angle at which it leaves the surface (the **angle of reflection**). This is the law of reflection. The dashed line drawn at 90° to the surface is called the **normal**. Angles are always measured from the normal and not from the surface.

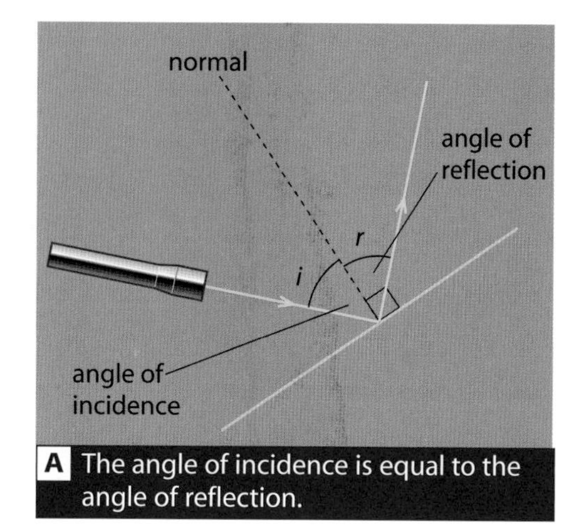

A The angle of incidence is equal to the angle of reflection.

When light passes from one transparent medium to another, for example from air into glass, some light is reflected and some passes through. As the light passes into the second material it changes speed and direction. When a wave experiences a change in direction due to a change in speed, this is **refraction**. Light travels in glass at two thirds of the speed it travels in air.

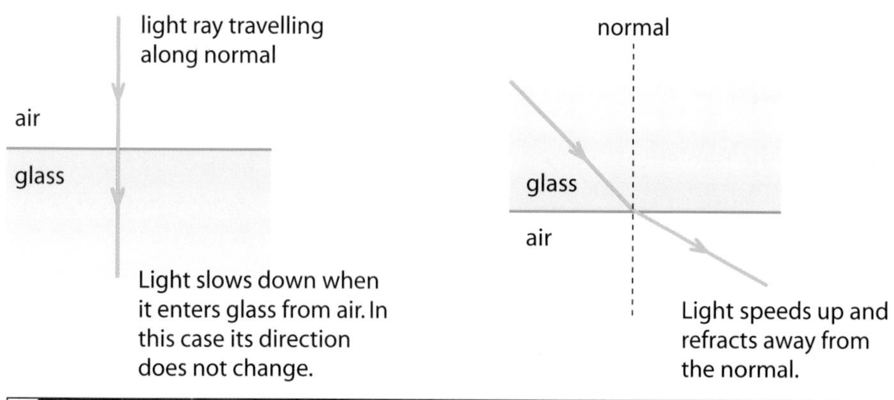

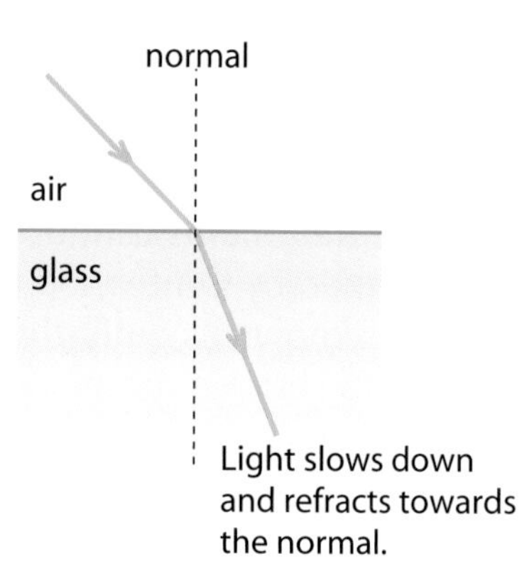

B Refraction of light rays passing between air and glass at different angles.

If light rays enter the new medium along the normal they change speed, but do not change direction. If light strikes the boundary at an angle there is a change of both speed and direction. When going from glass to air, the light rays speed up and refract away from the normal. When going from air to glass, the rays slow down and refract towards the normal.

When light arrives at a boundary between two materials, the light can be both reflected and refracted, or all of it can be reflected. It depends on the angle of incidence. If light moves from a denser material such as water to a less dense material such as air, most light is refracted away from the normal, but some light is reflected.

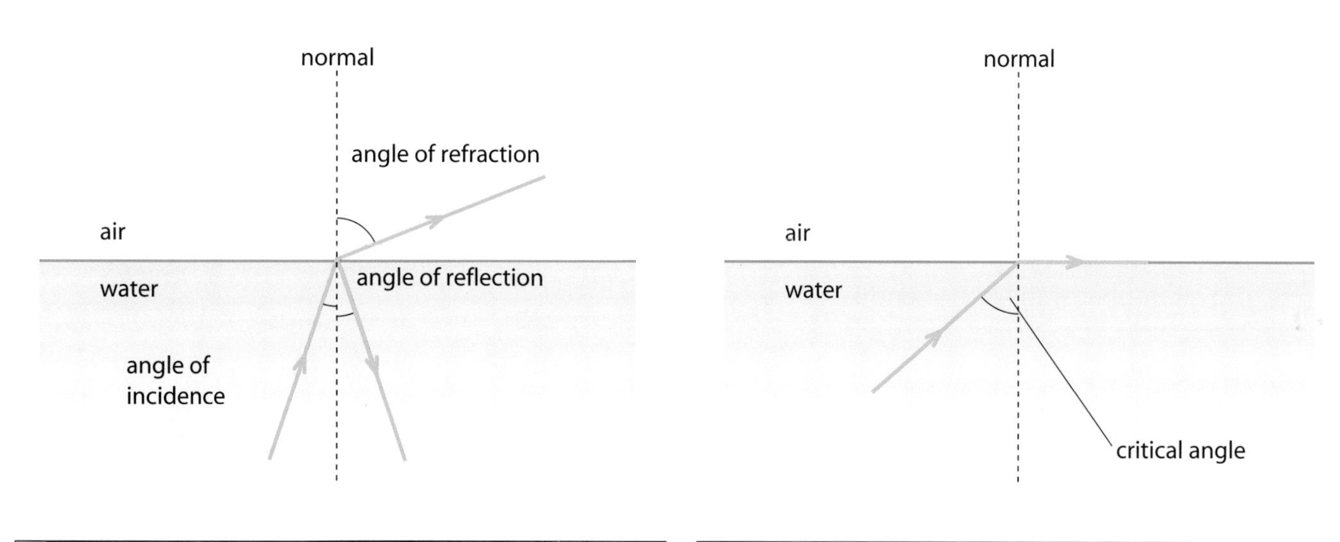

C Some light is reflected when the angle of incidence is low.

D Angle of incidence is at the critical angle.

If the angle of incidence increases to a certain angle, called the critical angle, then all the light is refracted at 90° to the normal, along the boundary.

1 What is:
- **a** an incident ray
- **b** a reflected ray
- **c** the critical angle?

2 When light travels from one piece of glass to another, identical piece of glass, it does not refract. Explain why.

If the angle of incidence is above the critical angle, then all of the light is reflected, staying inside the denser material. This is called **total internal reflection**.

3 What happens to light when it hits a boundary with a less dense material at an angle:
- **a** less than the critical angle
- **b** equal to the critical angle
- **c** greater than the critical angle?

4 Why is total internal reflection useful?

5 Draw four annotated diagrams to show light moving from air into glass
- **a** at 90° to the boundary
- **b** below the critical angle
- **c** at the critical angle
- **d** above the critical angle.

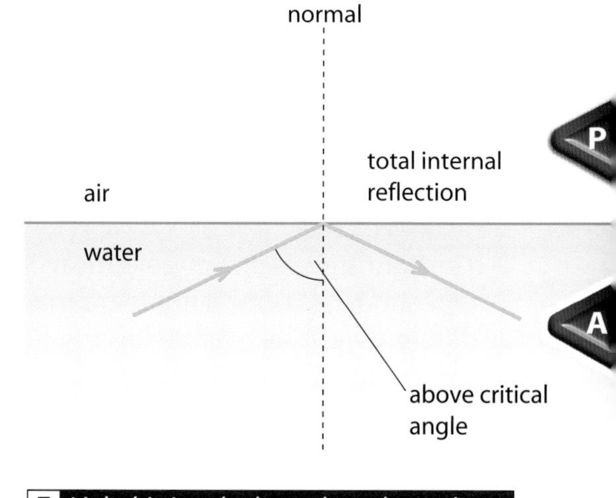

E Light hitting the boundary above the critical angle.

7. Endoscopy

By the end of these two pages you should be able to:

- explain the role of total internal reflection in the transmission of light along an optical fibre
- describe the use of optical fibres in endoscopes
- give examples of the use of endoscopes.

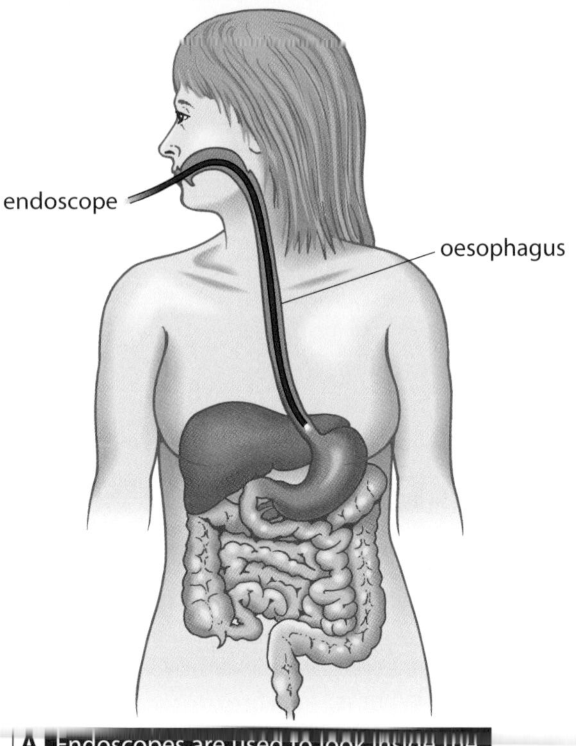

A Endoscopes are used to look inside the body.

Have you ever wondered?

How can you 'see' things inside the human body?

An **endoscope** is a slender tube which allows a doctor to look into the passageways of the body without having to operate. It can be inserted through the mouth or other natural openings. Alternatively a small cut can be made in the skin to allow inspection of the chest, abdomen or a joint, to take samples or to remove obstructions and diseased areas.

Endoscopes consist of a rigid or flexible tube containing glass fibres called **optical fibres.** The endoscope allows the **transmission of light** into and out of the body. It has a light source attached, and the light passes along one set of optical fibres, down the endoscope and out at the end. The light is reflected off the objects inside the body and then the light travels back up a different set of optical fibres to the eyepiece. The doctor looks at the image through the eyepiece, or it is displayed on a screen.

1 The inside of the body is dark. Briefly explain how light can get inside to allow the doctor to see.

2 The part of the endoscope which enters the body is long and thin. What other design features should an endoscope have to make it suitable for its job?

The light travels down the optical fibre by total internal reflection. The optical fibres are made of two layers of glass of different densities. On the inside there is high-density glass surrounded by an outer cladding, or coat, of low-density glass. The light hits the boundary between the two layers above the critical angle, and so is reflected back. This ensures that light entering one end will travel all the way to the other end without leaving the edges of the fibre by refraction. At the end of the endoscope the light is reflected off a surface inside the body; it then travels back through the glass fibre, each time being reflected from the boundary with the low-density glass.

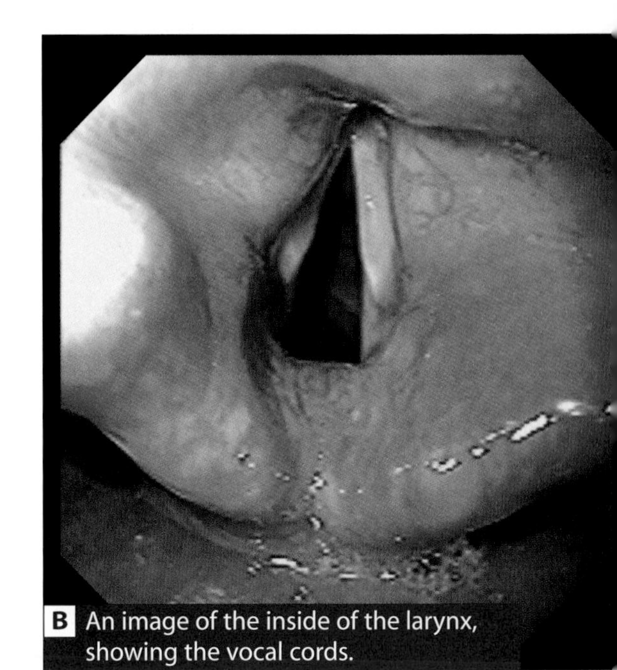

B An image of the inside of the larynx, showing the vocal cords.

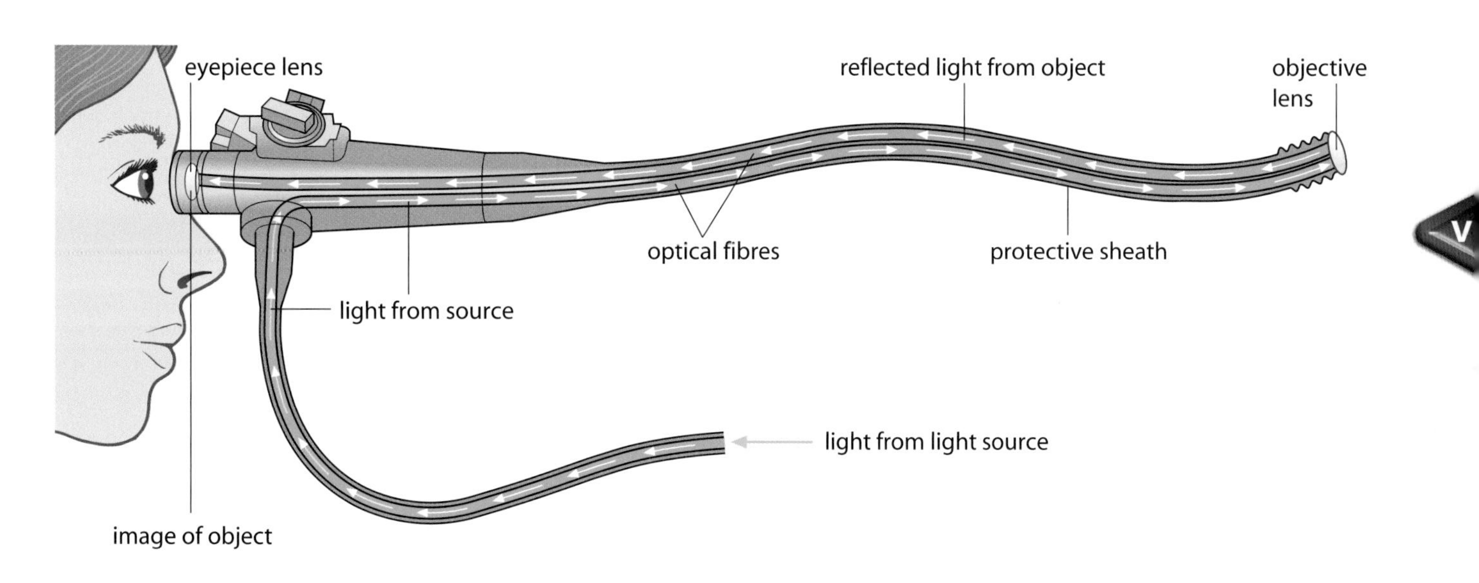

C The route taken by light in an endoscope.

Most endoscopes are flexible and movement is controlled by buttons next to the operator's eyepiece. On some, there are switches and controls to send in air or water or to suck out air. Some endoscopes have attachments which allow the doctor to remove pieces of tissue or to cauterise (burn) them.

Have you ever wondered?

Will 'seeing' these things hurt and is it safe?

Medical uses of endoscopes include:

- arthroscopy: looking inside and treating joints
- bronchoscopy: looking for tumours, abscesses and tuberculosis in the lungs
- colonoscopy: examination of the large intestine for ulcers and inflammation
- cystoscopy: used for the bladder, urethra and prostate
- gastroscopy: looking at the lining of the oesophagus, stomach and duodenum
- laparoscopy: an incision in the abdominal wall to see the fallopian tubes, stomach, liver or other organs.

3 Explain how total internal reflection works in optical fibres to produce an image of the inside of part of the body in the eyepiece.

4 Give three medical uses of endoscopy.

5 Design an endoscope that could be used to destroy kidney stones that have moved into the bladder. Describe how it works.

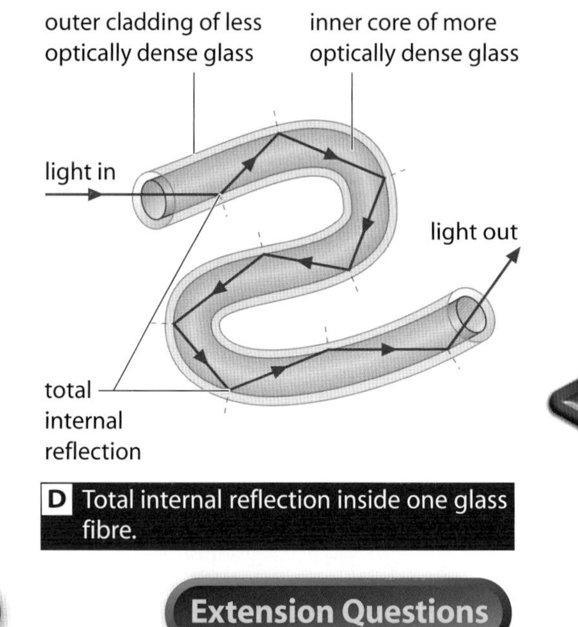

D Total internal reflection inside one glass fibre.

Summary Exercise

Higher Questions

Extension Questions

8. Introducing pulse oximetry

By the end of these two pages you should be able to:

- describe how pulse oximetry can be used to measure pulse rate.

Have you ever wondered?

How can scientists now check your blood is healthy without spilling any?

Your heart pumps oxygenated blood around the body. As the blood passes through the arteries, they expand and you can feel this as a pulse. If you put a finger against the artery at the wrist or in the neck, you can record the number of times your heart beats per minute – your pulse rate. On average your heart beats around 70 times a minute.

If you want to record the pulse rate more accurately, or over a longer period of time, you use a device called a pulse oximeter. This can measure two variables; your pulse rate and how much oxygen is in the blood.

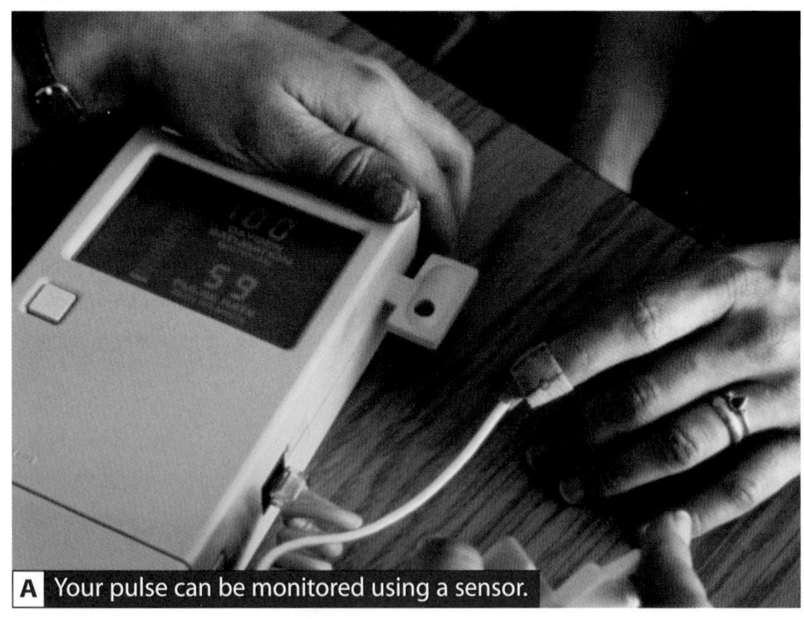

A Your pulse can be monitored using a sensor.

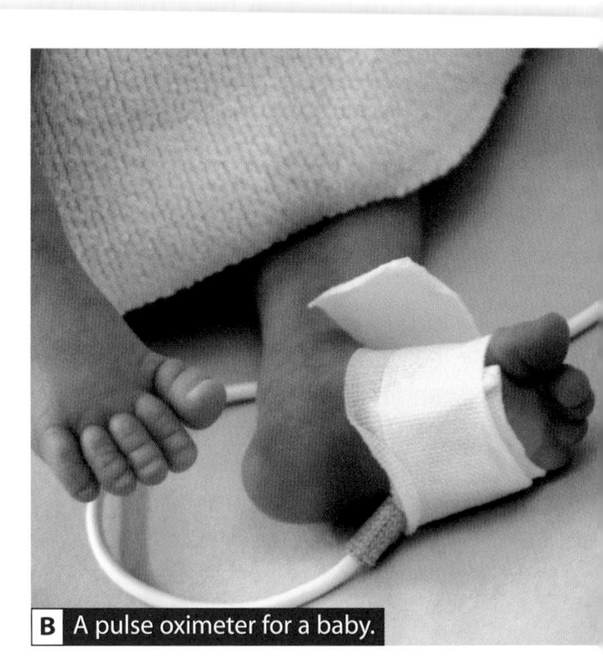

B A pulse oximeter for a baby.

Pulse oximetry is a simple, non-invasive and continuous method of monitoring your heart rate. It is often used for patients during an operation, when they are under anaesthetic. It can also be used to determine whether a person has a disease that affects the blood. Some pulse oximeters display waveforms of your pulse; some emit a sound if the pulse rate is too high or too low. Oximeters are also able to detect when a patient has too little oxygen (hypoxia) before they show any visible sign, such as lips turning blue.

A probe is placed on a patient's finger or ear lobe. For babies, probes are often attached to the foot. Some sensors are disposable and are used just once. Others are cleaned with alcohol solution ready for the next patient.

A pulse oximeter consists of light-emitting diodes (LEDs) and photodetectors. The LEDs produce beams of red and infrared light which pass through body tissue such as a finger and are then detected by the photodetectors which measure the intensity of the light. There are two types of pulse oximeter. The most common type has the LEDs on one side of the tissue and the photodetectors on the other side. An alternative type works by reflection – both the LEDs and the photodetectors are on the same side of the tissue. Light produced by the LEDs moves through the tissue and is then reflected and detected by the photodetectors.

1 What is the source of light in a pulse oximetry probe?

2 What parts of the body can be connected to probes?

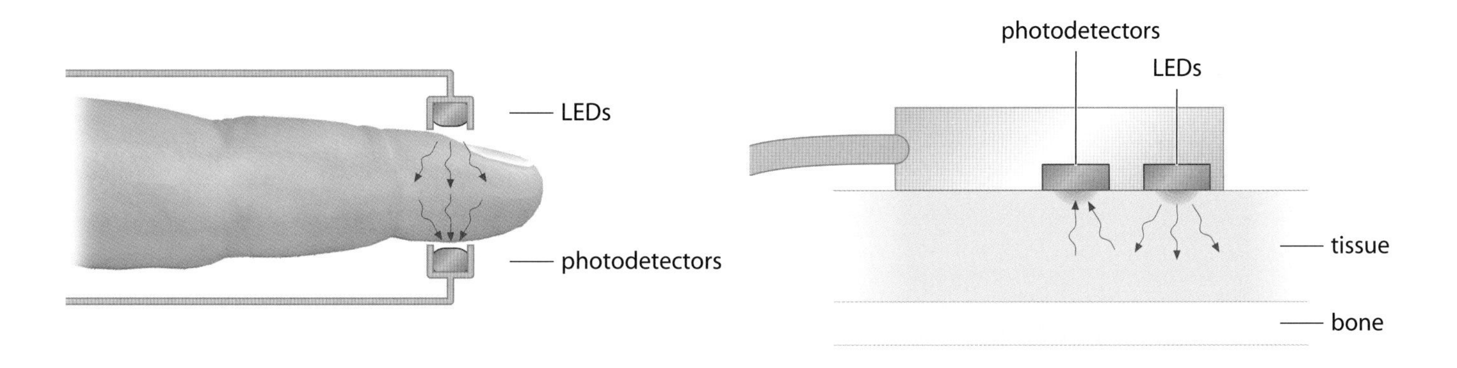

C Transmission and reflection pulse oximeters.

During a pulse the arteries stretch, their volume increasing in the moment after the heart contracts and pushes out blood. As the pulse of blood passes through the beam of light from the LEDs, the blood absorbs more of the light and the photodetectors receive less light, so the reading changes. Each change is detected as a pulse.

The LEDs do not shine continuously – they flash. During their off time, the pulse oximeter measures the light level in the room and allows for this when displaying a reading.

3 What causes the amount of light picked up by the photodetector to change with time?

4 When do you think it would be important to use disposable probes?

5 When would you use an oximeter which uses reflection, rather than the transmission type?

6 Draw annotated diagrams of a common pulse oximeter with the beam of red light when:

a a pulse of blood is passing underneath

b there is no pulse of blood.

Summary Exercise | Higher Questions | Extension Questions

9. Pulse oximetry and oxygen

By the end of these two pages you should be able to:

- describe how pulse oximetry can be used to monitor blood oxygen levels.

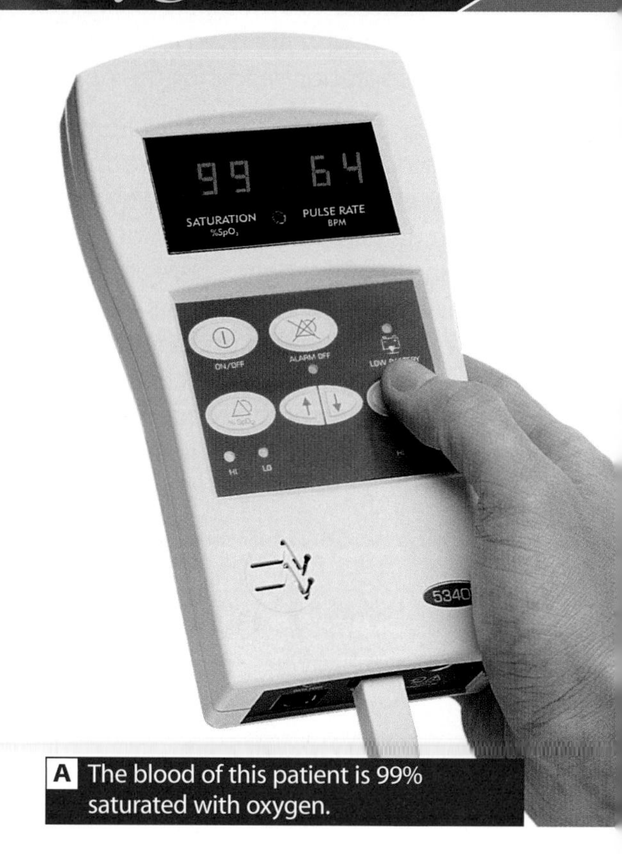

A The blood of this patient is 99% saturated with oxygen.

As well as measuring heart rate, pulse oximetry also monitors the oxygen level in the blood. Oxygen is needed by the millions of living cells in our body. It is carried in the blood, attached to haemoglobin molecules.

Oxygen saturation is a measure of how much oxygen the blood is carrying, compared to the maximum it could carry. One haemoglobin molecule can carry up to four atoms of oxygen. So one hundred haemoglobin molecules could carry a maximum of 400 oxygen atoms. If they were actually carrying 380 atoms of oxygen, they would be carrying $^{380}/_{400} \times 100\% =$ 95% of the maximum. The blood would be described as being 95% saturated. This is written 95 SpO_2% or 95 SPO2%.

The colour of blood depends upon the amount of oxygen the haemoglobin is carrying. Oxygenated blood, which has a high oxygen saturation, is a bright crimson red colour. Blood with a lower oxygen saturation is much darker.

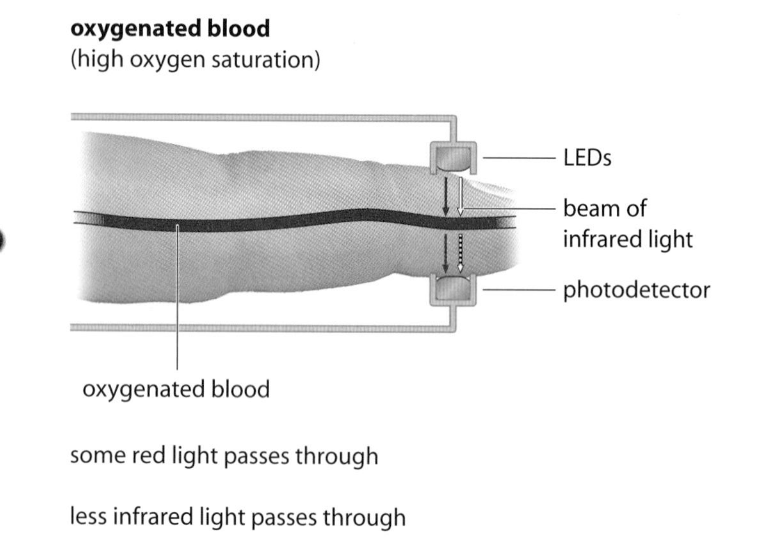

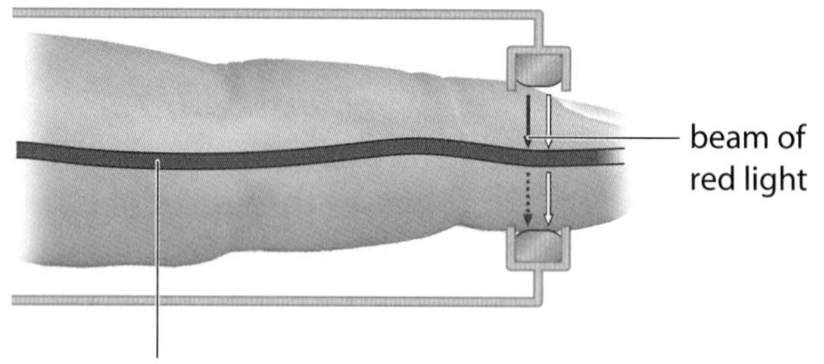

B How a pulse oximeter measures blood oxygen levels.

The light-emitting diodes (LEDs) in a pulse oximeter actually shine two beams of light through your finger: red light (which you can see), but also a beam of infrared light (which you cannot see).

The light from the LEDs is partly absorbed by haemoglobin, but not by the same amount for the red and infrared light. Oxygenated blood (blood saturated with oxygen) absorbs a lot of infrared light, but doesn't absorb so much red light. Poorly oxygenated blood (blood less saturated with oxygen) absorbs red light, but doesn't absorb as much infrared light. By calculating the difference in absorption of the red and the infrared beam, the machine can calculate the proportion of haemoglobin which is oxygenated.

Fingers are full of other materials which also absorb light. So to work out the colour of the arterial blood, a pulse oximeter looks for the slight change in the overall colour caused by the beat of the heart pushing blood through the artery.

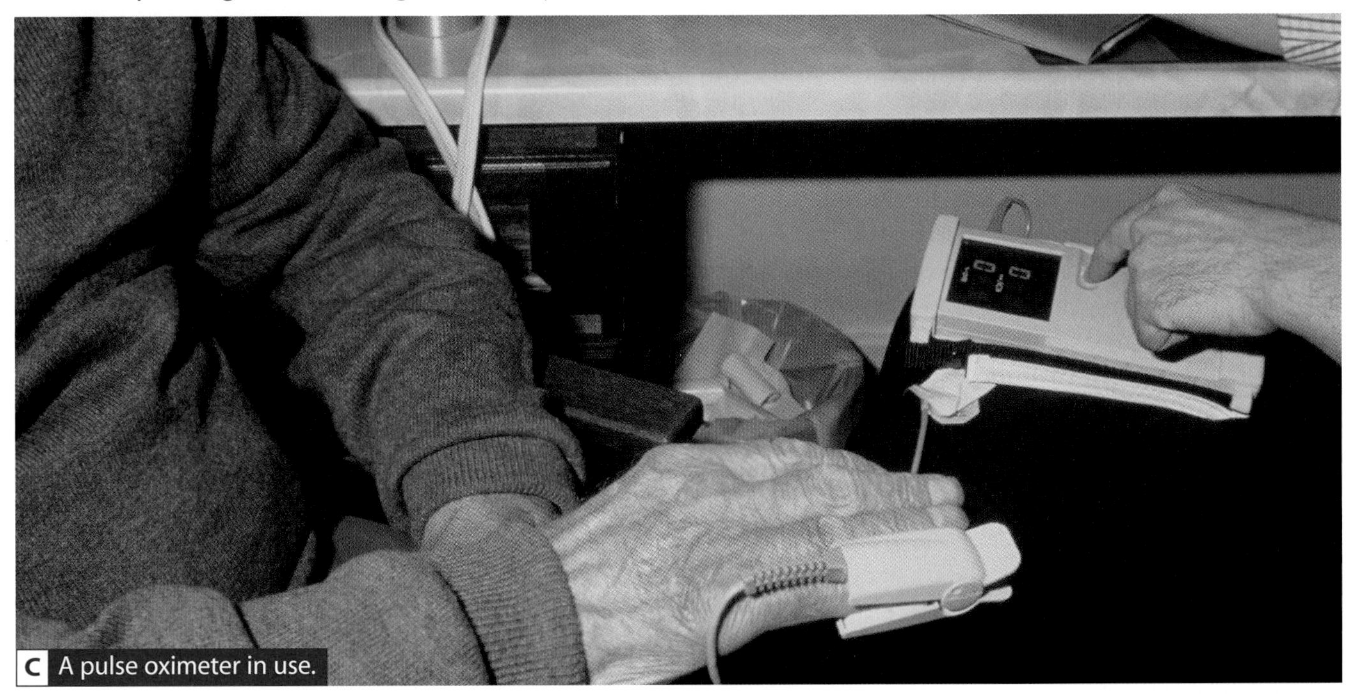

C A pulse oximeter in use.

1. What is the percentage saturation of blood which has 340 oxygen molecules for 100 molecules of haemoglobin?
2. In what way does oxygenated blood look different from blood which has given up some of its oxygen?
3. Why does the light coming from a probe appear to be just one colour, when there are two types of LED?
4. Explain how an oximeter measures oxygen saturation.

Summary Exercise

Higher Questions

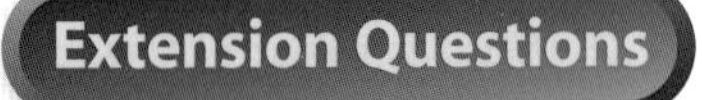

10. Heart electricity

By the end of these two pages you should be able to:

- recall that muscle cells can generate potential differences and that these can be used in medical applications
- describe how electrical potentials can be measured with an electrocardiogram (ECG) to monitor heart action
- explain the characteristic shape of a normal ECG in terms of heart action.

The heart is a pump. It pumps blood to the lungs and, at the same time, pumps blood rich in oxygen to all other parts of the body. It is made of cardiac muscle, famous for never tiring and being found nowhere else in the body. At rest, the heart contracts about 70 times a minute. When exercising, the contractions are stronger and the rate can double to 140 beats per minute, or more.

Have you ever wondered?

Does the human body produce electricity?

The heart has four chambers: two atria at the top and two ventricles at the base. Blood from the body enters the heart through the veins. It moves from the veins into the atria. The atria contract and push the blood down into the ventricles. Then the ventricles contract and push the blood out of the heart, through the arteries.

Each muscle cell must contract at the right moment. It is vital that a cardiac muscle cell contracts when its neighbouring cells are contracting. The instruction to contract is electrical and is generated within the heart. The signal that passes along a neurone is a change in its electrical potential and is called an **action potential**. Heart cells, unlike other cells, also transmit these electrical signals to instruct neighbouring cells to contract.

The electrical message which starts the contraction begins in a lump of specialised cardiac cells in one of the atria of the heart called the **sino-atrial node (SAN)**. The SAN is sometimes called the pacemaker. The SAN produces changes in the **potential difference** across the muscle cells. The change in voltage spreads across the atria as a wave of electrical **depolarisation** (change in charge) and causes the muscle cells in the atria to contract, pushing the blood downwards into the ventricles.

1 Which part of the heart produces the electrical signal which starts its contraction?

2 What is meant by depolarisation?

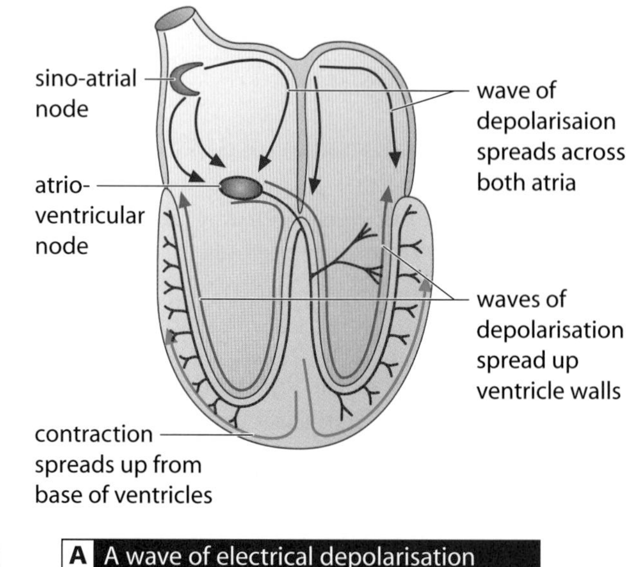

A A wave of electrical depolarisation causes heart muscle to contract.

Insulating fibres prevent the wave of depolarisation passing directly from the atria to the ventricles. Instead the electrical activity reaches the **atrio-ventricular node (AVN)**, in between the atria and the ventricles, which, after a short delay, sends the message down the wall between the ventricles. The wave of depolarisation reaches the bottom of the heart and spreads up the ventricle walls. This causes the contraction of the ventricles, starting at the bottom, squeezing the blood upwards to the arteries which carry the high-pressure blood away from the heart.

The human body is mainly salty water and so it conducts electricity. The electrical activity which spreads across the heart also spreads outwards away from the heart. This causes the voltage of the skin to change slightly. These potential differences are detected at various points on the body.

The picture of electrical signals produced by the heart, but picked up by skin, is called an **electrocardiogram**. This is commonly abbreviated to ECG (or EKG in the USA).

Diagram C shows an ECG, where each part of the wave is coloured to match the waves of depolarisation shown on the heart in diagram A.

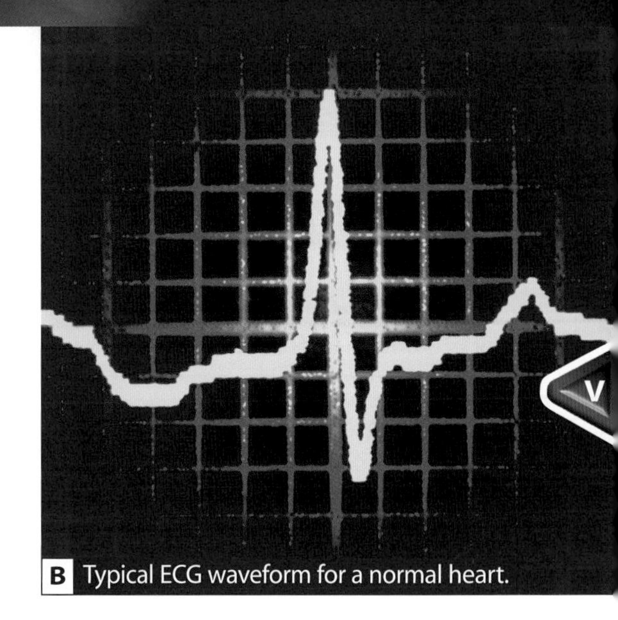

B Typical ECG waveform for a normal heart.

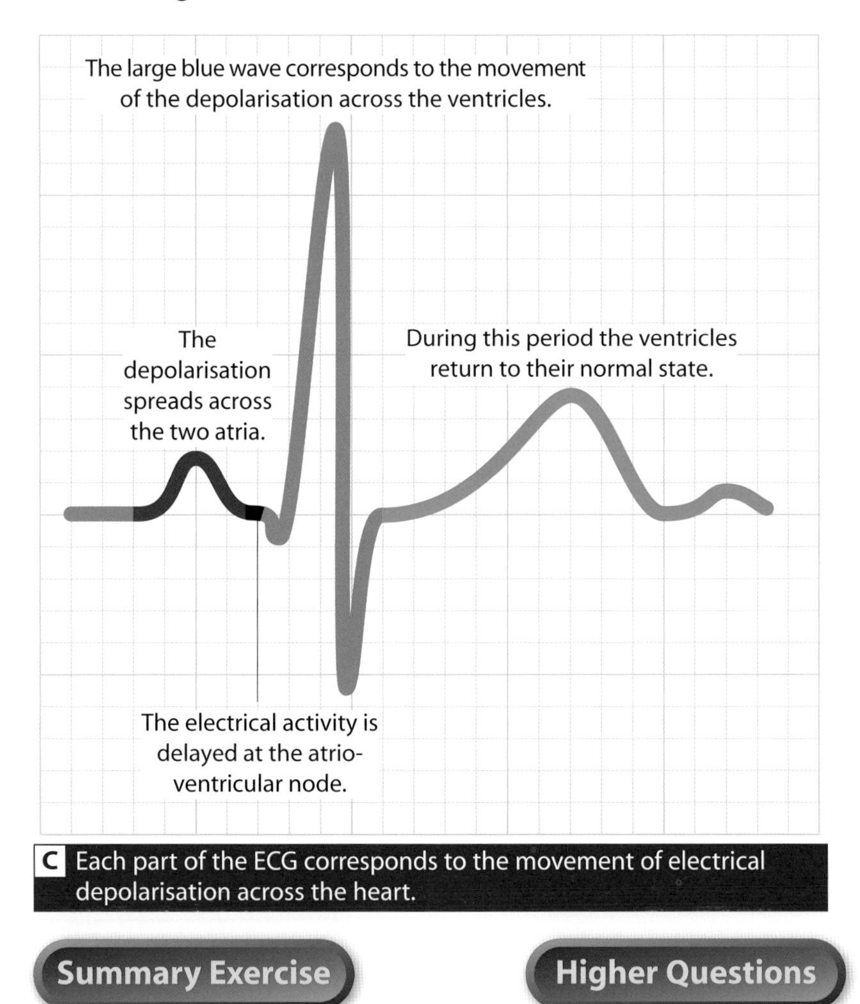

C Each part of the ECG corresponds to the movement of electrical depolarisation across the heart.

3 Where in the heart is the AVN situated?

4 If the skin did not conduct electricity, how else could you monitor the heart?

5 Annotate a sketch of a typical ECG to explain what is happening in the heart at each point.

Summary Exercise **Higher Questions**

Extension Questions

11. Reading ECGs

By the end of these two pages you should be able to:

- recall and use the equation: frequency = 1/time period, $f = 1/T$
- describe how changing electrical potentials can be measured with an ECG to monitor heart action.

Electrocardiograms (ECGs) are given labels to identify particular parts. The P wave occurs as the atria contract, the QRS complex is when the ventricles contract, and the T wave indicates that the ventricles are relaxing again. This whole process of one contraction and one relaxation of the heart is a cardiac cycle. If an ECG shows an unusual pattern, this shows that there is a problem with the heart.

ECGs are always printed on the same standard graph paper. The *x*-axis is time, with each small division equal to 40 milliseconds (ms). The large box is therefore 200 ms = 0.2 s. The *y*-axis is the recorded voltage, usually set at 5 mm per millivolt (mV).

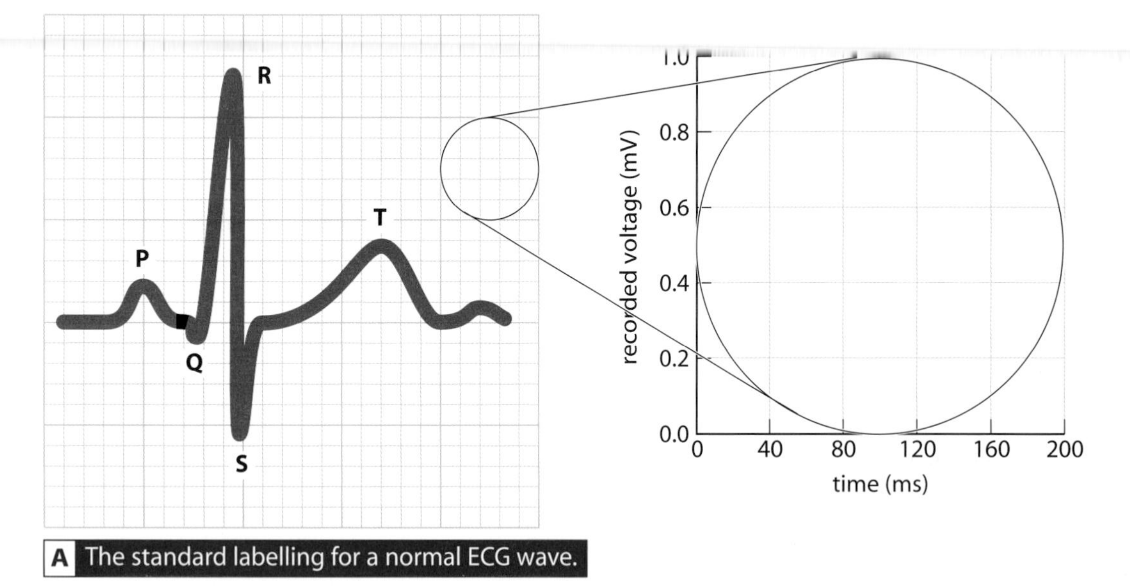

A The standard labelling for a normal ECG wave.

Diagram B shows an ECG from a healthy patient. It shows a regular beat and has the normal PQRST pattern.

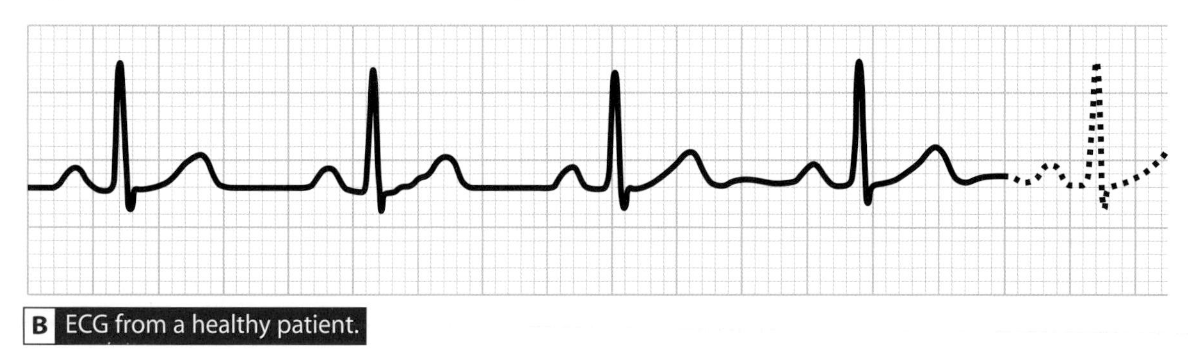

B ECG from a healthy patient.

You can use this ECG to find the patient's heart rate. First you need to count the number of complete cardiac cycles. In the example above there are four. Then you need to count how many large squares the trace of complete cycles covers. Here the four cardiac cycles cover 15 large squares. Remember that one large square represents 200 milliseconds (0.2 s).

Time for four heart beats = 15 boxes × 200 ms per large square
= 3000 ms
= 3 s

So there are four heart beats every 3 seconds.

Number of beats in 1 s = 4 beats/3 s = 1.33 beats/s
Number of beats in 1 min = 1.33 × 60 = 80 beats per minute (bpm)
The heart rate is therefore 80 bpm.

Frequency is defined as the number of cycles per second. We can calculate frequency using the equation:

$$\text{frequency}, f = \frac{1}{\text{time period}, T}$$

Looking at the same ECG, the length of one cardiac cycle can be taken as the distance from the top of one P wave to the top of the next P wave. This distance is 3.75 large squares. This length is the same if measured from any point on the wave to the corresponding point on the next wave, i.e. from one S to the next S.

Time for one beat = 3.75 large squares × 200 ms per box
= 750 ms
= 0.75 s

So the time for one beat is 0.75 s. This is the period, the length of time for one cardiac cycle. The symbol for period is T. For this patient, $T = 0.75$ s.

frequency = 1/time period = 1/0.75 s = 1.33 beats per second

Note that 1.33 beats per second is the same as 80 bpm. It is usually more accurate to measure the distance for several cycles, rather than for just one cycle.

4 Draw two ECGs. Both should show normal PQRST waveforms, but one should be for a heart beating too slowly and the other should be beating too fast.

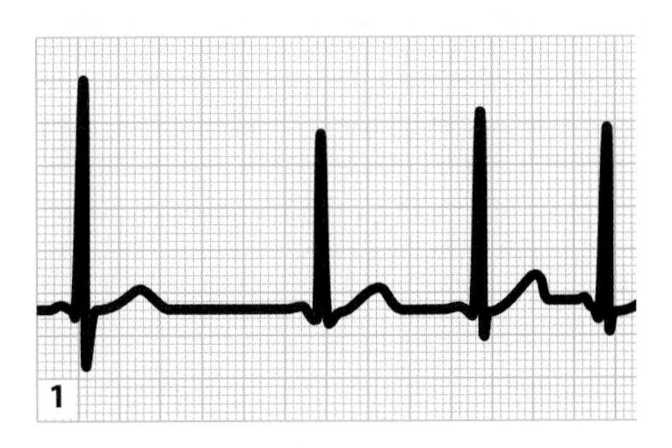

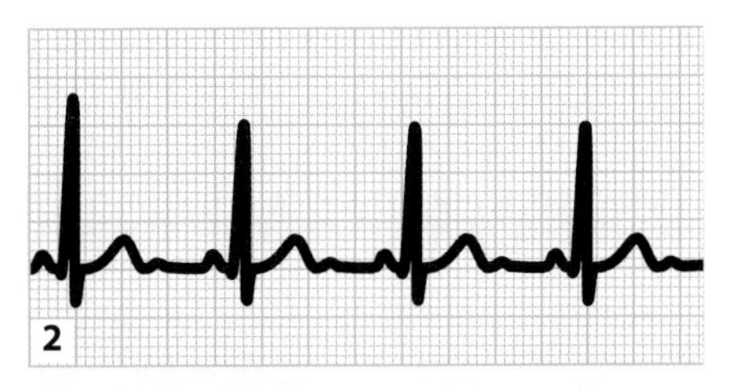

1 Look at ECG 1.
- **a** What is the shortest period shown on this ECG?
- **b** What is the longest period shown on this ECG?

2 Look at ECG 2.
- **a** What is the heart rate of this patient, in beats per minute?
- **b** What is the period for one cardiac cycle?

3 Do both ECGs show normal PQRST waves?

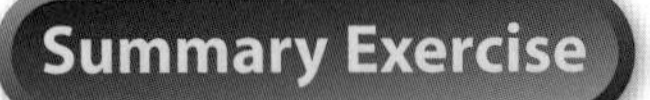

Higher Questions

12. PET scans

By the end of these two pages you should be able to:

- describe examples of the use of positron emission tomography (PET) scanning.

Have you ever wondered?

Can you look at a medical image and tell whether your body is working properly?

For many decades, ionising radiations such as X-rays have been used to produce images of the inside of the body. A traditional X-ray produces a shadow image when passing through dense objects in the body. These images are useful for dentistry and to show broken bones, but to see small items in the body much more detail is needed. **Computerised tomography (CT)** and **magnetic resonance imaging (MRI)** use computer imaging, rather than a picture on a film, and are able to make images which show slices of the body (**tomography**) with superb clarity. CT uses X-rays and MRI uses a strong magnetic field.

Positron emission tomography (PET) scans are quite different to other medical imaging techniques because they do not display the anatomy of the body. Instead PET scans show the position of body processes. A PET scan reveals areas with high levels of chemical activity, such as high glucose metabolism. The cells in a cancerous tumour are dividing rapidly, so have high levels of glucose metabolism. The tumour then shows up on a scan as a dark region.

Often the PET scan is combined with a CT image to give a combined PET/CT image. The 'hot spots' detected by the PET are overlaid onto the sliced CT X-ray image to locate the areas of high metabolism precisely.

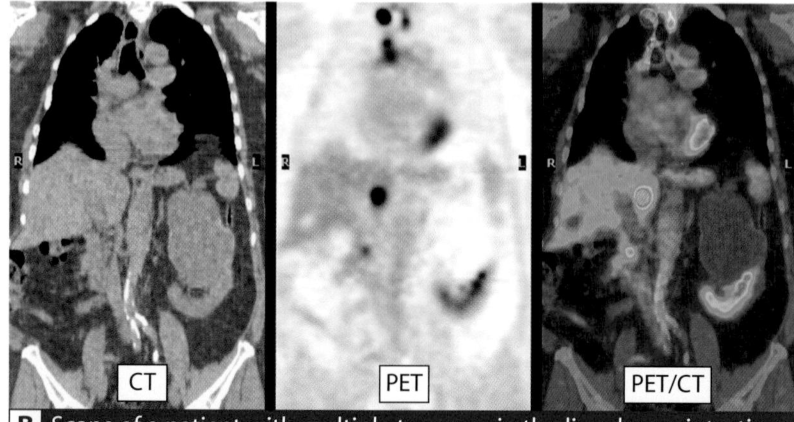

B Scans of a patient with multiple tumours, in the liver, lungs, intestines and nervous system.

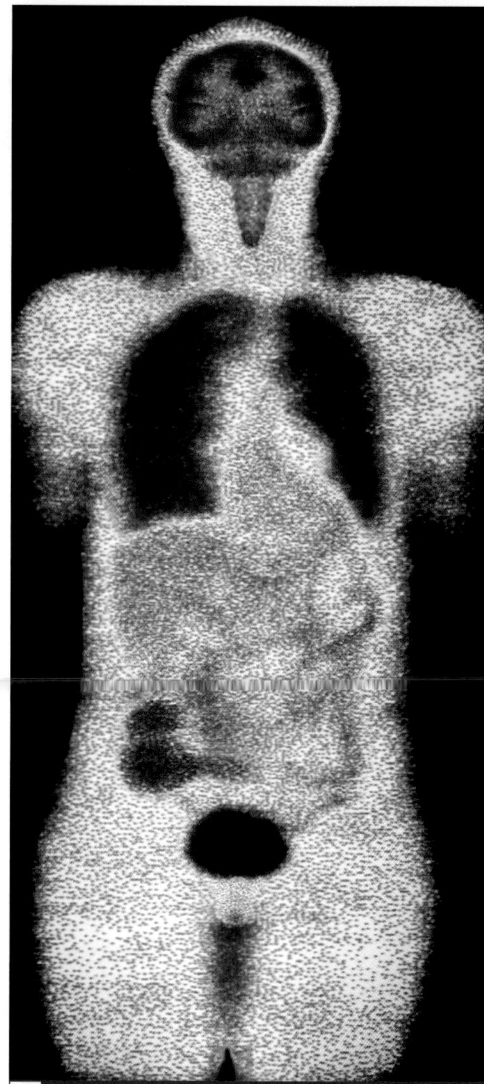

A This PET scan shows the position of a colon cancer. The brain also shows up because it has a high level of glucose metabolism.

1 What is tomography?

2 Which types of medical investigation use tomography?

PET works using a drug called a **radiopharmaceutical**, or radioactive tracer. This drug is made by combining a chemical which will travel to certain parts of the body, together with a radioactive element. The most common tracer is glucose combined with fluorine-18 and is called FDG (2-fluoro-2-deoxy-D-glucose). Fluorine-18 is radioactive and decays by emitting a positron. It has a half-life of about 110 minutes. The FDG collects in areas of high glucose metabolism and gives off radiation.

The radiopharmaceutical is injected into the arm of the patient and in less than an hour it will have collected in certain parts of the body. For instance, the brain uses glucose as its main source of energy. Glucose will go to those parts of the brain which are most active.

The radiopharmaceutical emits a type of radiation called **positrons**. Positrons are positively charged particles with the same mass as electrons – they are the **anti-particles** of electrons. The positrons collide with electrons in the body and produce gamma rays which travel out of the body. The level of radiation is similar to that from a normal X-ray.

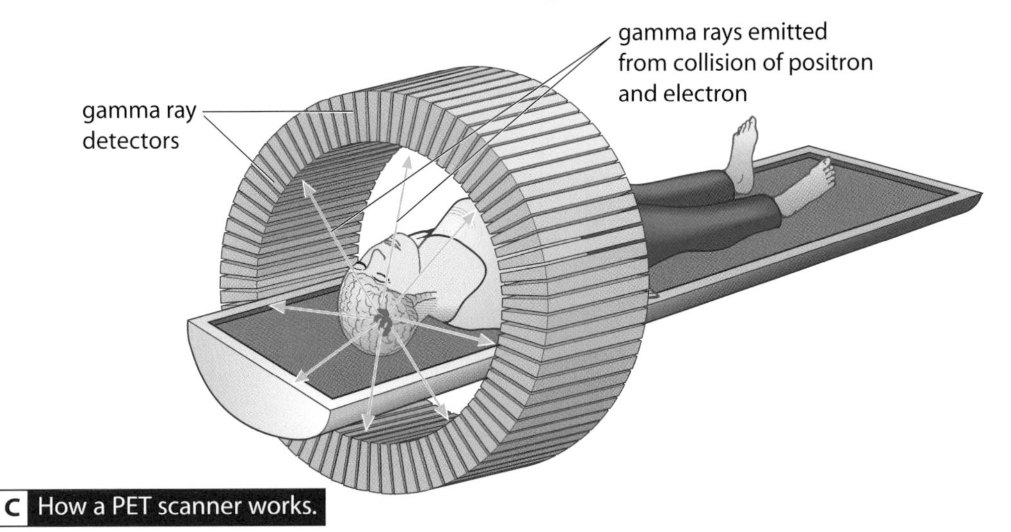

C How a PET scanner works.

The patient lies on a bed, surrounded by a ring of gamma ray detectors. The bed can be moved through the ring.

Any gamma rays passing from the patient through the ring of sensors are detected. Since each sensor points in one direction only, the computer knows which line the gamma ray travelled along. The computer produces a picture showing the position of the source of the radiation.

The bed moves slightly to a new position. The gamma radiation emitted from this plane is now detected and a new slice is displayed.

3 What is a radiopharmaceutical?

4 Should patients undergoing a PET scan be concerned about becoming radioactive?

5 Why are PET scans often combined with CT scans?

6 The radiology department of your hospital wants a sign for the door to explain what PET scans are. Design a sign using no more than 80 words.

Summary Exercise **Higher Questions**

13. Bombardment

By the end of these two pages you should be able to:

- describe what is meant by a thermal neutron
- use thermal neutrons in nuclear equations
- describe the bombardment of certain stable elements with proton radiation, to make them into radioactive isotopes that usually emit positrons.

The radiopharmaceutical used in PET scanning releases positrons. The radiopharmaceutical is a radioactive **isotope** which can be produced by **bombarding** a stable element with proton radiation.

In order to bombard an element with protons, or other charged particles, the protons need to be moving at high speed and energy. This can be done by accelerating them in a type of particle accelerator called a cyclotron.

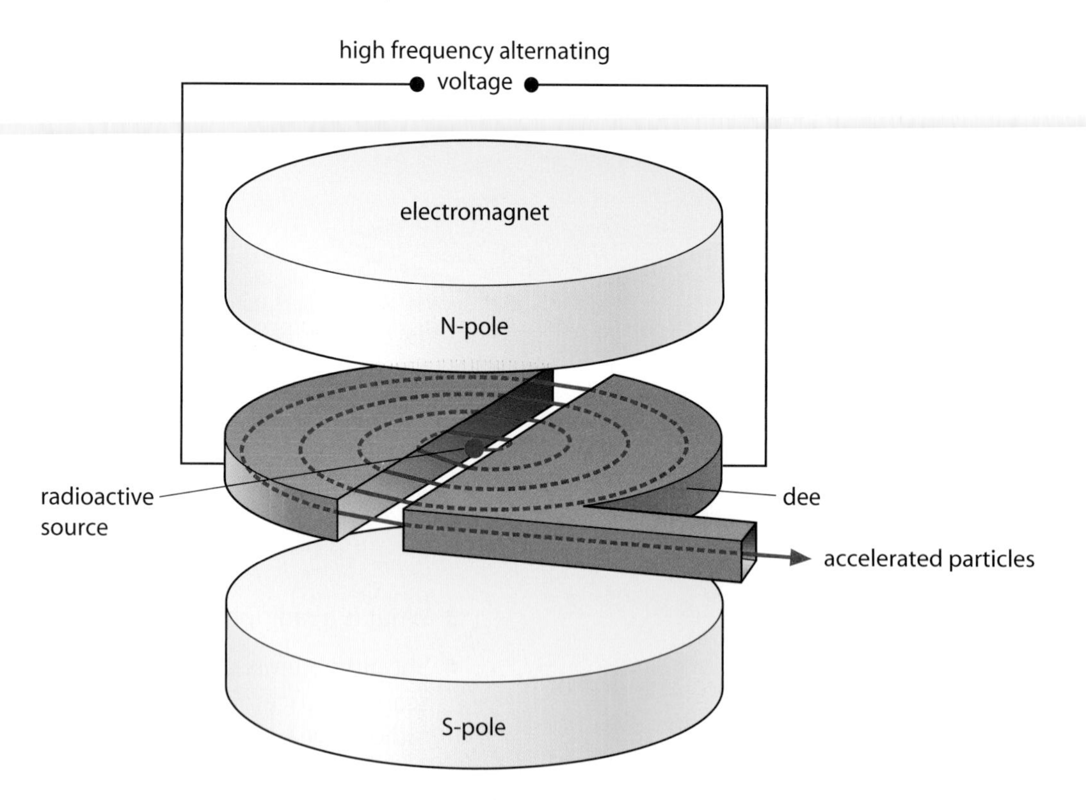

A A cyclotron accelerates charged particles.

A cyclotron consists of two electromagnets, which create a magnetic field in the region between them. Inside this magnetic field are two electrodes called D-electrodes, or dees, because they are D-shaped. The dees are hollow so that particles can move inside them, and there is a small gap between them. A radioactive source in the centre, between the dees, releases protons or other charged particles.

A voltage is applied to one of the dees, to attract the particles. Once the particles are inside the dee, the magnetic field forces the protons to move in a circular path. As the particles leave the first dee, the voltage on the second dee is changed so that it attracts the particle. A positive particle, for example, will accelerate towards the next dee if it is more negative. The particles gain speed while they are in the gap between the electrodes. So the particles move in a spiral, getting faster each time they pass between the dees, until they leave through the exit tube.

In a cyclotron, accelerated protons are used to bombard stable atoms to produce the positron-emitting radioisotopes used in PET scanning.

Some of the cyclotron-produced radioisotopes used in PET scanning are very short-lived and must be administered to the patient very soon after production. For this reason, cyclotrons are installed next to the PET scanners.

1 What is a positron?

2 How does a cyclotron produce a fast-moving particle?

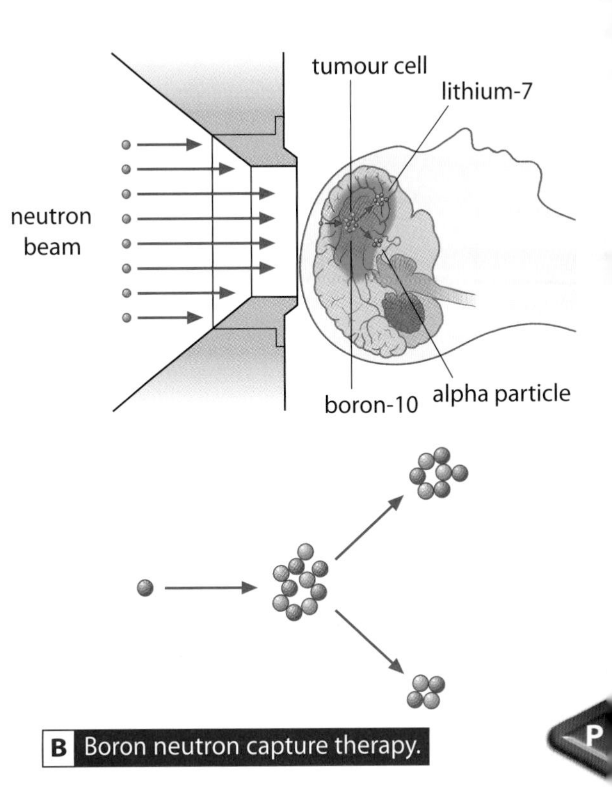

B Boron neutron capture therapy.

Another way of producing radioisotopes is in nuclear reactors using **thermal neutrons**. These are slow, low-energy neutrons which have a kinetic energy similar to that of the surrounding particles of gas. They are free neutrons, not bound in the nucleus of an atom, and are useful in chain reactions. Stable atoms can be bombarded with thermal neutrons to create a heavier isotope of the atom. This new isotope is neutron-rich and the excess of neutrons makes the nucleus unstable – it is radioactive.

Thermal neutrons are used in boron neutron capture therapy (BNCT), where a stable isotope of boron-10 is attached to tumour cells inside the body. The tumour cells are then bombarded with a beam of thermal neutrons. The boron captures the neutron, then undergoes fission to form lithium and an alpha particle. These particles destroy any cells nearby – the cancer cells – but do not penetrate further, so healthy cells are not affected.

$${}^{1}_{0}n + {}^{10}_{5}B \rightarrow {}^{11}_{5}B$$

$${}^{11}_{5}B \rightarrow {}^{7}_{3}Li + {}^{4}_{2}\alpha$$

BCNT should therefore reduce side-effects of other cancer treatments such as radiotherapy and chemotherapy, and is particularly useful for brain tumours.

3 What is a thermal neutron?

4 What are two uses of thermal neutron bombardment?

5 What is BNCT?

6 Make a concept map of bombardment, showing different types and uses.

Summary Exercise

Higher Questions

Extension Questions

14. Annihilation

By the end of these two pages you should be able to:

- describe what happens when a positron meets with an electron (they annihilate each other, with the production of gamma rays)
- explain that the meeting of a positron with an electron is an example of mass/energy conservation.

Whenever a particle meets its anti-particle they are **annihilated** immediately. The two particles are destroyed and all mass is converted to new particles and/or radiant energy. In positron emission tomography (PET) a positron collides with an electron, the two particles are annihilated and both are changed into gamma radiation.

The equation is given below. Positrons can be shown in three ways:

β^+ or ${}^{0}_{+1}\beta$ or ${}^{0}_{+1}e$.

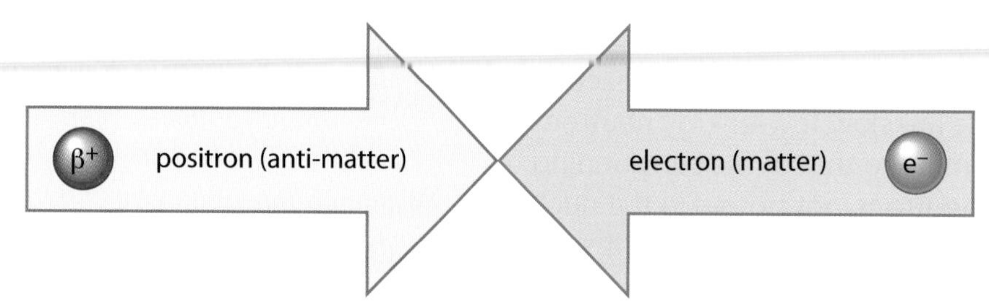

The positron and the electron have opposite charges so they attract each other.

They collide and annihilate (they are totally destroyed).

Gamma radiation spreads away from the point of annihilation.

A When matter and anti-matter collide, the particles are annihilated.

The annihilation that takes place in a PET scanner is:

$${}^{0}_{+1}e^+ + {}^{0}_{-1}e^- \rightarrow 2\,{}^{0}_{0}\gamma$$

positron + electron → two gamma rays each of 8.2×10^{-14} J

The gamma rays produced are detected by the PET scanner.

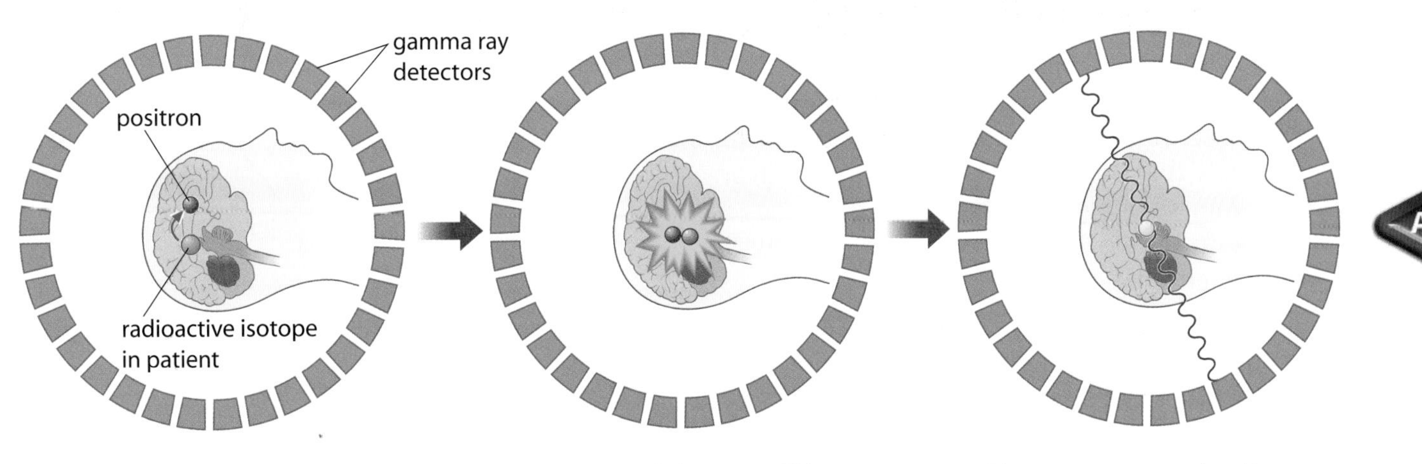

Radioactive isotope emits positron.

Positron and electron collide, annihilating each other.

Two gamma rays emitted in opposite directions and picked up by detectors.

B The PET scanner detects gamma rays.

Annihilation occurs when any type of particle collides with its anti-particle, and in all cases the total of mass and energy is conserved. The law of **energy conservation** states that energy cannot be created or destroyed – it can only be transferred from one type to another. So the total energy before the collision is the same as the total energy after it.

1 What is annihilation?

2 Describe how the process of annihilation is used in PET scanners.

When a proton and an electron with negligible kinetic energy annihilate each other, they produce two identical gamma rays. The two original particles have the same mass (9.11×10^{-31} kg each). Einstein's famous equation, $E = mc^2$, can be used to calculate total energy from the conversion of mass.

Before the collision:

energy, E, in joules = mass, m, in kilograms × (speed of light, c, in metres per second)^2

$= 2 \times 9.11 \times 10^{-31}\ \text{kg} \times (3.00 \times 10^8\ \text{m/s})^2$

$= 1.64 \times 10^{-13}\ \text{J}$

After the collision, two gamma rays are produced. The energy of each gamma ray is 8.2×10^{-14} J. As there are two rays, the total energy after the collision is 1.64×10^{-13} J.

Hence energy (in the form of mass or electromagnetic radiation) is conserved.

3 What is the law of energy conservation?

4 How is positron–electron annihilation an example of mass/energy conservation? Use the annihilation equation to explain your answer.

Summary Exercise

Higher Questions

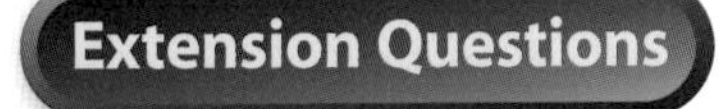

15. Momentum

By the end of these two pages you should be able to:

- show that the meeting of a positron with an electron is an example of momentum conservation
- perform calculations on momentum conservation in one dimension.

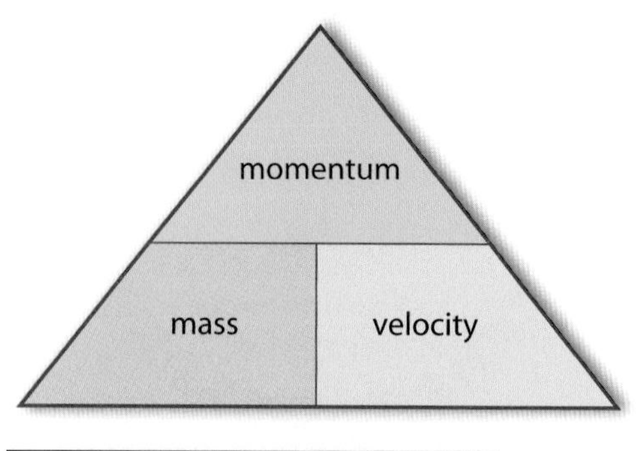

A Momentum memory triangle.

Momentum is a property of moving objects. Calculations using momentum can be used to predict the behaviour of things, big or small, that are colliding or exploding.

The momentum of an object can be calculated using the equation:

momentum, ρ (kgm/s) = mass, m (kg) × velocity, v (m/s)

The momentum of an object is its mass multiplied by its **velocity**. As the term velocity involves both size and direction, momentum also involves both size and direction; it is an example of a **vector** quantity. It is measured in kilogram metres per second.

In all these examples, movement to the right will be positive.

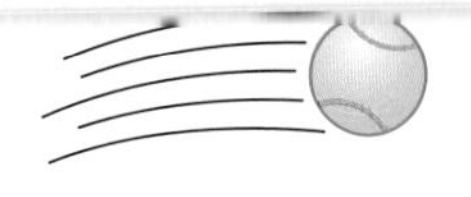

v = +16 m/s
m = 100 g
momentum = 0.1 kg × +16 m/s
= +1.6 kgm/s

v = +7 m/s
m = 40 kg
momentum = 40 kg × +7 m/s
= +280 kgm/s

v = +24 m/s
m = 1000 kg
momentum = 1000 kg × +24 m/s
= +24 000 kgm/s

B Momentum examples.

The law of **momentum conservation** (in a straight line) says that the total momentum before a collision is equal to the total momentum afterwards.

A small car runs into the back of a slow-moving larger car. Immediately after the collision the smaller car is travelling at only +8m/s, whereas the other car is now travelling at +15m/s.

v = +20 m/s
m = 1000 kg
momentum = 1000 kg × +20 m/s
= +20 000 kgm/s

v = +5 m/s
m = 1200 kg
momentum = 1 200 kg × +5m/s
= +6 000 kgm/s

Total momentum before collision = +20 000 kgm/s (small car) + +6 000 kgm/s (large car)
= +26 000 kgm/s

C Momentum of cars before collision.

$v = +8$ m/s
$m = 1000$ kg
momentum = 1000 kg × +8 m/s
= +8000 kgm/s

$v = +15$ m/s
$m = 1200$ kg
momentum = 1 200 kg × +15 m/s
= +18 000 kgm/s

Total momentum after collision = +8 000 kgm/s (small car) + +18 000 kgm/s (large car)
= +26 000 kgm/s

D Momentum of cars after collision.

The total momentum is conserved: no momentum is lost or gained during the collision.

1 What is the momentum of a football of 0.42 kg travelling at a velocity of +16 m/s?

2 A car of 1200 kg travelling at +20 m/s hits a stationary car of 800 kg. The two stick together. What is their velocity after collision?

The law of conservation of momentum also applies to the collision of particles, including positron–electron annihilation in PET scanners.

Inside the patient, the positron given off by the radiopharmaceutical is attracted to an electron and collides with it. Since they are travelling in opposite directions, one of them will have a negative velocity and a negative momentum, the other will have a positive velocity and a positive momentum. A positive and a negative momentum of the same amount cancel out, so before collision the total momentum of the two particles is zero.

After collision, if only one gamma ray was produced, then it would have momentum, which would break the law. However, two gamma rays are produced and travel in opposite directions. So one gamma ray has a negative momentum relative to the other. This means that their momentums equal zero. So, the total momentum before and after the positron–electron annihilation is the same. Momentum is conserved.

If one or both of the original positron or electron have a significant velocity at collision, then there is some momentum to start with. In this case, the gamma rays produced do not move in opposite directions. The PET scanner detects that they are not 180° apart, and ignores them.

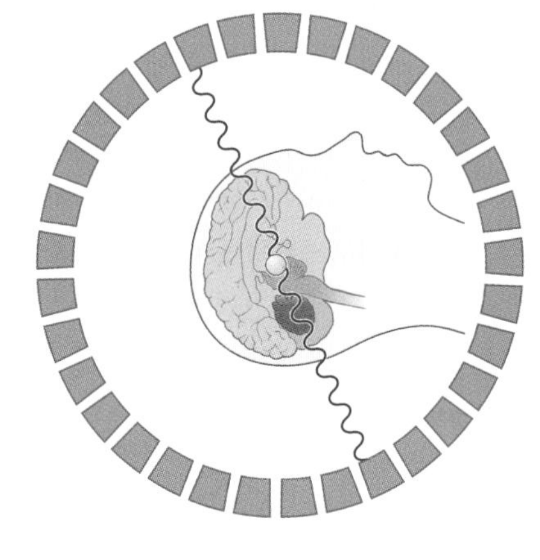

Two gamma rays emitted in opposite directions and picked up by detectors.

E The two gamma rays produced by an electron and a positron annihilating each other have equal and opposite momentums.

3 Why is the total momentum of the positron and electron zero before they collide?

4 Describe how electron–positron annihilation in a PET scanner shows conservation of momentum.

Summary Exercise
Higher Questions
Extension Questions

16. Ethical issues

By the end of these two pages you should be able to:

- describe some social and ethical issues relating to the introduction of new medical techniques.

Dear Mrs Adams,

Your family doctor, Dr D. K. Patel, has referred your daughter Ann to the Medical Physics and Radiology Department of this hospital. In order to confirm whether Ann's lump is cancerous or not, we propose carrying out a PET scan. This involves injecting a small amount of a radioactive liquid into her arm. An hour after the injection, she will be asked to lie on a bed surrounded by a scanner. Radiation coming from the radioactive liquid in her body will be detected for up to 30 minutes and displayed on a screen for a radiologist to make a diagnosis.

As Ann has reached puberty, to reduce any risks to her reproductive organs we would like to use a new radiopharmaceutical. This is currently under phase II clinical trials. Experiments on rats and on healthy paid volunteers have been very positive and well within limits of acceptable harm. I can assure you that I feel that the use of this tracer is best for Ann and poses no more risk than two traditional X-rays.

Yours sincerely

Dr H. B. Stocker
Consultant Radiologist

A Ann and her mother.

1 Why does the doctor want Ann to have a PET scan?

2 How might some radiopharmaceuticals affect Ann's reproductive organs?

3 The letter asks Ann to use a drug that is in phase II clinical trials.
  a What are clinical trials?
  b What would be the advantages and the disadvantages to Ann of using the new type of drug?

4 Do you agree with clinical trials? Should we test new drugs or medical techniques on humans? Give reasons for your answer.

5 Should Ann have the PET scan, and should she use the new drug? Give reasons for your answer.

Dear Dr Kehinde,

I am delighted to inform you that your bid for special funding has been successful. The Department of Health has awarded you the full amount of £8.2 million over 5 years for your new PET scanner. This will cover all costs for land acquisition, build, capital purchase, installation, staff training, staff salaries and daily operating expenses.

You must build a centre operating a combined whole-body PET/CT scanner with a cyclotron to produce the radioisotopes. You guarantee to provide a minimum of 1400 whole body scans each year. This facility is for exclusive NHS use and is not to be used for private patients or research.

Can I use this opportunity to remind you that the Department's bidding scheme is attracting negative coverage in the media at the moment. The press will point out that for this sum of money your Trust could provide kidney dialysis for 60 patients for 5 years or provide thousands of other treatments.

We trust that you will be ready to explain the need for a PET scanner.

Yours sincerely,

Dr Richard Dent

B Dr Kehinde.

6 Is it right for that such a large sum of money be spent on a PET scanner rather than offer the other treatments mentioned in the letter? Give reasons for your answer.

7 The companies building the PET centre and supplying the equipment will make a lot of money. Should the government force them to lower their prices? Explain why.

8 Should the senior doctors in the health authority also work as directors of medical companies? Explain your answer.

17. Questions

D

P

1 Nav works in a shop. He lifts a box of chocolate bars with a weight of 250 N through a height of 1.8 m.
  a Calculate the work he does against gravity.
  b Each bar of chocolate provides 220 calories (kcal) of energy. What is the energy content of one bar expressed in joules? (1 calorie (kcal) = 4200 J)
  c If Nav ate one bar and all of that energy was converted to work, how many boxes could he lift through a height of 1.8 m?

2 When Ali is asleep, her body needs energy at a rate of 110 J/s.
  a Her basal metabolic rate is 80 J/s. Give one reason why the energy she needs while sleeping is higher than her basal metabolic rate.
  b She walks to her job at a speed of 1.8 m/s and experiences an average frictional force of 60 N. How much work does she do against friction in 5 minutes?

3 One of the pieces of equipment to be checked each morning in an operating theatre is the pulse oximeter.
  a Why does an oximeter have two types of LED?
  b Explain how an oximeter is able to calculate the percentage oxygen saturation of blood.

4 a What causes the change in voltage shown on an ECG?
  b On a particular ECG, the time taken for four cardiac cycles is 5.6 s.
    i What is the period for one cycle?
    ii What is the frequency expressed in beats per second?
    iii What is the frequency expressed in beats per minute?

5 PET scans are useful for medical diagnosis and scientific research.
  a What does PET stand for?
  b What is a radiopharmaceutical?
  c What is a positron?
  d What happens when a positron meets an electron?

6 A large car with a mass of 1200 kg travelling at 24 m/s collides with a stationary car with a mass of 1000 kg.
  a What is the law used to predict the behaviour of moving objects in collisions and explosions?
  b If the two cars stick together, what is their combined velocity after collision?

7 a What is radiotherapy?
  b A beam of X-rays is adjusted so that its intensity 20 cm from a point source is 40 mW/cm^2. What will be the intensity 40 cm from the source?
  c At this point, the radiation enters a patient. What will happen to the intensity as the X-rays pass through the patient's body?

8 a Name a type of radiation which is ionising and is produced by the decay of an unstable nucleus.
  b Name a type of radiation which is electromagnetic, non-radioactive and does not have enough energy to knock electrons out of the atoms of the medium it is passing through.
  c What are the usual units for the intensity of an electromagnetic wave?

9 After a serious incident where radioactive material was released, a policewoman is found to have received a radiation dose of 860 mSv.
  a What is this dose expressed in Sv?
  b What could she do now to keep her exposure to ionising radiations as low as possible? Name three things.

10 You live in a low-radon area where the annual radioactivity is 2 mSv. For two months you stay in Cornwall where the annual radiation dose is 6 mSv higher. A single chest X-ray gives you a dose of 0.02 mSv. How many X-rays would you have to have, to be equivalent to the extra radiation due to your holiday?

11 a Describe the structure of an endoscope.
  b With reference to critical angle, reflection and refraction, explain how it is possible for a beam of light to remain inside an optical fibre which has high-density glass surrounded by a cladding of lower-density glass.

12 The football manager said, "I have the power to make you work with energy until we win".

a Explain the scientific meanings of *power*, *work* and *energy*.

b The average power of a professional footballer is 680 W. How much energy is converted during a match lasting 90 minutes?

c Whilst exercising, the footballer converts food and stored energy to other types of energy. Name three types of energy produced whilst playing sport.

13 A high intensity beam of alpha particles is a potentially dangerous ionising radiation.

a Explain the terms *intensity*, *ionising* and *radiation*.

When passing through a vacuum the intensity of gamma radiation is 18 mW/m^2 at 15 cm from a point source.

b Express 18mW/m^2 in W/m^2.

c What will be the intensity 45 cm from the point source?

14 A hospital leaflet contains this sentence: "For malignant cancer the treatment will probably start with radiotherapy. If a tumour is benign, radiotherapy is sometimes used to provide palliative care."

a Explain the terms *benign*, *radiotherapy* and *palliative care*.

b Describe the effects of excessive exposure of the human body to ionising radiations.

15 Light is used in two common diagnostic devices: a pulse oximeter and an endoscope.

a How does the light in a pulse oximeter differ from light used in an endoscope?

b Explain how an oximeter measures pulse rate.

c An endoscope contains bundles of very thin optical fibres. How does light stay in one optic fibre?

d Describe one medical use of an endoscope.

16 An ECG detects the depolarisation of the heart, which begins at the sino-atrial node.

a Explain the terms *ECG*, *depolarisation* and *sino-atrial node*.

An ECG shows 12 complete PQRST waves in 8 seconds.

b What is the period of one cardiac cycle?

c What is the frequency, in beats per second?

d What is the heart rate, in beats per minute?

17 A PET scan detects gamma rays produced when annihilation occurs when a positron meets its anti-particle.

a Explain the terms *gamma ray* and *annihilation*.

b What is the name and charge of the anti-particle to a positron?

Positrons are emitted by a radiopharmaceutical injected into the patient.

c What factors would be considered when choosing a certain radiopharmaceutical for a patient?

d How does a cyclotron accelerate a positively charged particle?

18. Glossary

absorbed dose The amount of radiation energy received per kilogram.

*__action potential__ Change in voltage across a neurone or the membrane of a cardiac muscle cell as an electrical impulse travels along it.

angle of incidence The angle between the normal and the wave when it hits a surface.

angle of reflection The angle between the normal and the wave when it leaves the surface.

*__annihilate__ To destroy so that only energy remains.

anti-particle Has the opposite charge but same mass as its corresponding particle, which it annihilates on contact.

atrio-ventricular node (AVN) Area of cardiac muscle which starts the contraction of the lower part of the heart.

*__basal metabolic rate (BMR)__ The minimum amount of energy needed when you are at rest.

benign tumour A lump of cells which is not cancerous.

*__bombardment__ To fire a particle at the nucleus of an atom.

computerised tomography (CT or CAT) An X-ray picture which shows a slice through the body.

critical angle When incident light meets a boundary at the critical angle, the angle of refraction is 90°.

depolarisation Change in voltage.

*__electrocardiogram (ECG)__ A graph showing the change in voltage produced by the heart.

*__endoscope__ A tube which can be inserted into the body which allows the doctor or operator to see inside the body.

*__energy conservation__ Whenever energy is transferred from one type to another, there is no loss or gain in the total amount of energy.

*__gamma rays__ A form of electromagnetic radiation which has the highest frequency in the electromagnetic spectrum.

*__intensity__ The strength of a wave defined as power of incident radiation/area.

ion A particle, atom or group of atoms which has a positive or negative charge.

ionisation When charged particles are produced.

ionising radiation Radiation that can cause charged particles to be formed.

isotopes Atoms of the same element having a different number of neutrons in their nucleus – e.g. carbon-12 (six protons, six neutrons); carbon-14 (six protons, eight neutrons).

magnetic resonance imaging (MRI) This uses magnets to produce detailed images of slices through the body.

malignant tumour A cancerous growth.

*__momentum conservation__ In all collisions and explosions the total amount of momentum remains the same.

normal An imaginary line which is at right angles to a surface that a wave is hitting.

*__optical fibre__ Tight bundle of glass fibres which transmits light from one end to the other.

*__palliative care__ Treatment intended to ease suffering although it may not cure the illness.

*__positron__ The anti-particle of an electron, having the same mass but opposite charge. Positrons are emitted from isotopes with too few neutrons.

*__positron emission tomography (PET)__ A medical scanning technique. The image of metabolically active sites inside the body is computed by detecting gamma rays coming from positron–electron annihilation.

*__potential difference__ The voltage between two points.

*__power__ The amount of energy converted each second.

*__pulse oximetry__ Using a pulse oximeter to measure pulse rate and oxygen level of the blood.

*__radiation__ The emission of energy from a source, as sound, electromagnetic waves or moving particles.

radioactive When the unstable nucleus of an atom decays.

radiopharmaceutical A chemical which has been made radioactive.

reflection When a wave or particle bounces off.

*__refraction__ When light changes direction due to a change in speed.

sievert Unit used for the harm caused by a radiation.

sino-atrial node (SAN) Area of cardiac muscle which triggers the electrical stimulation of the heart.

*__source__ Something that gives out energy.

*__thermal neutron__ A slow-moving neutron.

tomography A process which produces an image that is a slice through the body.

*__total internal reflection__ When all of a wave is reflected from a boundary.

*__transmission of light__ The passage of light.

*__tumour__ An abnormal growth of rapidly dividing cells.

velocity How fast an object is moving in a certain direction.

vector A quantity which has size and direction.

*__work__ Energy converted when movement is opposed by a force.

*glossary words from the specification

Periodic Table

relative atomic mass — 1
H
hydrogen
atomic number — 1

Period	Group 1	2											3	4	5	6	7	0
1																		4 He helium 2
2	7 Li lithium 3	9 Be beryllium 4											11 B boron 5	12 C carbon 6	14 N nitrogen 7	16 O oxygen 8	19 F fluorine 9	20 Ne neon 10
3	23 Na sodium 11	24 Mg magnesium 12											27 Al aluminium 13	28 Si silicon 14	31 P phosphorus 15	32 S sulphur 16	35.5 Cl chlorine 17	40 Ar argon 18
4	39 K potassium 19	40 Ca calcium 20	45 Sc scandium 21	48 Ti titanium 22	51 V vanadium 23	52 Cr chromium 24	55 Mn manganese 25	56 Fe iron 26	59 Co cobalt 27	59 Ni nickel 28	63.5 Cu copper 29	65 Zn zinc 30	70 Ga gallium 31	73 Ge germanium 32	75 As arsenic 33	79 Se selenium 34	80 Br bromine 35	84 Kr krypton 36
5	85 Rb rubidium 37	88 Sr strontium 38	89 Y yttrium 39	91 Zr zirconium 40	93 Nb niobium 41	96 Mo molybdenum 42	(98) Tc technetium 43	101 Ru ruthenium 44	103 Rh rhodium 45	106 Pd palladium 46	108 Ag silver 47	112 Cd cadmium 48	115 In indium 49	119 Sn tin 50	122 Sb antimony 51	128 Te tellurium 52	127 I iodine 53	131 Xe xenon 54
6	133 Cs caesium 55	137 Ba barium 56	139 La lanthanum 57	178 Hf hafnium 72	181 Ta tantalum 73	184 W tungsten 74	186 Re rhenium 75	190 Os osmium 76	192 Ir iridium 77	195 Pt platinum 78	197 Au gold 79	201 Hg mercury 80	204 Tl thallium 81	207 Pb lead 82	209 Bi bismuth 83	(209) Po polonium 84	(210) At astatine 85	(222) Rn radon 86
7	(223) Fr francium 87	(226) Ra radium 88	(227) Ac actinium 89	(261) Rf rutherfordium 104	(262) Db dubnium 105	(266) Sg seaborgium 106	(264) Bh bohrium 107	(277) Hs hassium 108	(268) Mt meitnerium 109	(271) Ds darmstadtium 110	(272) Rg roentgenium 111							

Index

Edexcel
190 High Holborn
London WC1V 7BH
UK

Fourth impression 2008

ISBN: 978-1-846901-53-9

Designed by Roarrdesign
Illustrated by Oxford Designers and Illustrators Ltd
Picture research Charlotte Lippmann
Indexer John Holmes
Printed and bound in China GCC/04

The publisher's policy is to use paper manufactured from sustainable forests.

Acknowledgements

The Publisher would like to thank Rachel Brown, Head of Science, The Mirfield Free Grammar and Sixth Form; Ian Round, Head of Science, Monk's Walk School; Dr. Mike O'Neill, Director of Educational Strategies, UCST & ULT and Mike Viccary, Head of Science, Bradford Christian School for their help in the production of this book.

We are grateful to the following for permission to reproduce photographs:

(Key: b-bottom; c-centre; l-left; r-right; t-top)

Alamy Images: Andrew Lambert/LGPL 104, 121; D.Hurst 107; G.P.Bowater 106; Images of Africa/David Keith Jones 67t; Jennie Hart 193tl; Nick Gregory 22b; Siegfried Kuttig 193tr; Zefa RF 64t; Ardea: Steve Hopkin 52; Benecol: 11br; Bridgeman Art Library Ltd: Burrell Collection, Glasgow, Scotland 69; CERN Geneva: 153, 153tl, 184tl, 184tr, 185; Trevor Clifford: 81tl, 84b, 86b, 89, 90bl, 90br, 90t, 91b, 91t, 93t, 94b, 95b, 97, 101, 110, 112t, 118b, 120, 122tl, 122tr, 124, 126b, 128l, 128r, 130, 131, 133cl, 133t, 134b, 136, 138b, 140, 141b, 147t, 148, 149t, 1394.1134, 1412t; Corbis: Andrew Fox 74; Bio Sidus/Handout/Reuters 41; EPA/KCNA 18t; Kazuyoshi Nomachi 19; Peter Dench 23; Reuters 40; Courtsey of Kimbolton Fireworks: 85t; Ecoscene: Sally Morgan 142; FLPA Images of Nature: Cyril Ruoso/Jh Editorial/Minden 68; Foto Natura/Flip de Nooyer 56t; Foto Natura/Jan Vermeer 56b; Frans Lanting/Minden 47, 58; Gerry Ellis/Minden 59tl; Jurgen and Christine Bohns 76; Matthias Breiter/Minden 62t; Ron Austing 62b; Sunset 57; Winfried Wisniewski 54; Food Features: 11t, 14, 15; Getty Images: AFP/Odd Andersen 77; Digital Vision 17; Mario Tama 24; Time Life/NIna Leen 50; Iain Brand: 27t, 109, 111b, 112b; iStock Photo: 20; JIm Clark: 103; Johnson Matthey: 119; Kikkoman Trading Europe GmbH: 13t; Lebrecht Music and Arts Photo Library: Wladimir Polak 118t; Mari Tudor-Jones: 9t, 12, 18, 25, 45tl, 45tr, 48t, 55b, 92t, 117tl, 117tr, 138t, 147c; Paul Mulcahy: 144; Nature Picture Library: Andy Sands 61cl; E.A. Kuttapan 63; Jane Burton 65; NHPA Ltd / Photoshot Holdings: Andy Rouse 64b; Jean-Louis Le Moigne 61tr; Steve Robinson 59tr, 72b; PMS Instruments: 206; Reckitt Benckiser plc: 149b; Rex Features: SIPA Press 73; Science Photo Library Ltd: 55t, 134t, 158, 207; AJ Photo 204cr; Alfred Pasieka 29, 209; Alison Wright 31; Andrew Lambert 85bl, 162; Andy Crump,TDR.WHO 34b; ArSciMed 182; Biophoto Associate 38; Brookhaven National Laboratory 181; Charles D. Winters 84t; Cordelia Molloy 189; Andy Crump 34t; Custom Medical Photo/Keith 83b; David M. Martin 202; David Nunuk 35; Dr Morley Read 32r; Dr. Jeremy Burgess 126t; Ed Young 1892tr; Francoise Sauze 174; Geoff Tompkinson 170; ISM 212b; James Prince 204cl; Karsten Schneider 154; Martin Stankewitz 32; Martyn F. Chillmaid 102; Maximilian Stock Ltd 130b; Michael Gilbert 168; National Library of Medicine 173; Paul Rapson 111t; Professor Peter Goddard 67b; Scott Bauer/US Department of Agriculture 39; Scott Camazine 166; Shelia Terry 28; Simon Fraser 30; Simon Fraser,Main X-Ray, Newcastle General Hospital 167; Simon Fraser/Royal Victoria Infirmary,Newcastle Upon Tyne 37; Soverign, ISM 212tr; Stevie Grand 198; Thomas Porrett 164; Tony McConnell 157; Walter Dann 49; Shailesh M Shenoy and Robert H Singer from Figure2A, Science 297:836 (2002): 27bl; Pictures Courtesy of Southern Water: 81tr, 82, 83t, 86t, 113; STILL Pictures The Whole Earth Photo Library: Martin Harvey 60; Wellcome Trust Medical Photographic Library: 48b; WHO: P.Virot 33l; With thanks to A. Alfuraih and N.M.Spyrou, University of Surrey: 215

Front cover photo and back cover shot: © Photodisc / Getty Images

The Publisher is grateful to all the copyright owners whose material appears in this book.

Every effort has been made to trace the copyright holders and we apologise in advance for any unintentional omissions. We would be pleased to insert the appropriate acknowledgement in any subsequent edition of this publication.

Licence Agreement: *Edexcel 360Science Extension Units ActiveBook*

Warning:
This is a legally binding agreement between You (the user) and Edexcel Limited, 190 High Holborn, London WC1V 7BH, United Kingdom ('Edexcel Ltd').

By retaining this Licence, any software media or accompanying written materials or carrying out any of the permitted activities, You are agreeing to be bound by the terms and conditions of this Licence. If You do not agree to the terms and conditions of this Licence, do not continue to use the Disk and promptly return the entire publication (this Licence and all software, written materials, packaging and any other component received with it with Your sales receipt to Your supplier for a full refund.

Edexcel 360Science Extension Units ActiveBook consists of copyright software and data. The copyright is owned by Edexcel Ltd. You only own the disk on which the software is supplied. If You do not continue to do only what You are allowed to do as contained in this Licence you will be in breach of the Licence and Edexcel Ltd shall have the right to terminate this Licence by written notice and take action to recover from you any damages suffered by Edexcel Ltd as a result of your breach.

Yes, You can:

1. use *Edexcel 360Science Extension Units ActiveBook* on your own personal computer as a single individual user;

No, You cannot:

1. copy *Edexcel 360Science Extension Units ActiveBook* (other than making one copy for back-up purposes);
2. alter *Edexcel 360Science Extension Units ActiveBook*, or in any way reverse engineer, decompile or create a derivative product from the contents of the database or any software included in it;
3. include any software data from *Edexcel 360Science Extension Units ActiveBook* in any other product or software materials;
4. rent, hire, lend or sell *Edexcel 360Science Extension Units ActiveBook*;
5. copy any part of the documentation except where specifically indicated otherwise;
6. use the software in any way not specified above without the prior written consent of Edexcel Ltd.

Grant of Licence:
Edexcel Ltd grants You, provided You only do what is allowed under the Yes, 'You can' table above, and do nothing under the 'No, You cannot' table above, a non-exclusive, non-transferable Licence to use *Edexcel 360Science Extension Units ActiveBook*.

The above terms and conditions of this Licence become operative when using *Edexcel 360Science Extension Units ActiveBook*.

Limited Warranty:
Edexcel Ltd warrants that the disk or CD-ROM on which the software is supplied is free from defects in material and workmanship in normal use for ninety (90) days from the date You receive it. This warranty is limited to You and is not transferable.

This limited warranty is void if any damage has resulted from accident, abuse, misapplication, service or modification by someone other than Edexcel Ltd. In no event shall Edexcel Ltd be liable for any damages whatsoever arising out of installation of the software, even if advised of the possibility of such damages. Edexcel Ltd will not be liable for any loss or damage of any nature suffered by any party as a result of reliance upon or reproduction of any errors in the content of the publication.

Edexcel Ltd does not warrant that the functions of the software meet Your requirements or that the media is compatible with any computer system on which it is used or that the operation of the software will be unlimited or error free. You assume responsibility for selecting the software to achieve Your intended results and for the installation of, the use of and the results obtained from the software.

Edexcel Ltd shall not be liable for any loss or damage of any kind (except for personal injury or death) arising from the use of *Edexcel 360Science Extension Units ActiveBook* or from errors, deficiencies or faults therein, whether such loss or damage is caused by negligence or otherwise.

The entire liability of Edexcel Ltd and your only remedy shall be replacement free of charge of the components that do not meet this warranty.

No information or advice (oral, written or otherwise) given by Edexcel Ltd or Edexcel Ltd's agents shall create a warranty or in any way increase the scope of this warranty.

To the extent the law permits, Edexcel Ltd disclaims all other warranties, either express or implied, including by way of example and not limitation, warranties of merchantability and fitness for a particular purpose in respect of *Edexcel 360Science Extension Units ActiveBook*.

Governing Law:
This Licence will be governed and construed in accordance with English law.